EPIC OF THE EARTH

EPIC

OF THE

EARTH

Reading Homer's *Iliad* in the
Fight for a Dying World

EDITH HALL

Yale
UNIVERSITY PRESS
New Haven and London

Published with assistance from the Louis Stern Memorial Fund.

Yale University Press books may be purchased in quantity for
educational, business, or promotional use. For information, please
e-mail sales.press@yale.edu (U.S. office) or sales@yaleup.co.uk
(U.K. office).

Set in Janson type by Integrated Publishing Solutions.
Printed in the United States of America.

Library of Congress Control Number: 2024940442
ISBN 978-0-300-27558-2 (hardcover : alk. paper)

A catalogue record for this book is available from the British Library.

This paper meets the requirements of ANSI/NISO Z39.48-1992
(Permanence of Paper).

10 9 8 7 6 5 4 3 2 1

*For my children and the rest of Generation Z
with apologies for the state of the planet*

Contents

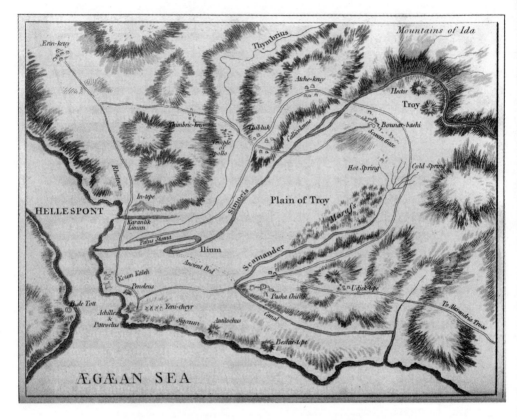

Map of Troy (Reproduced from James Dallaway and
Gaetano Mercati, *Constantinople Ancient and Modern*
[London: T. Bensley, T. Cadell, W. Davies, 1897])

EPIC OF THE EARTH

Prologue

THE EPIC POEM THE *Iliad* introduces the earliest detailed account of the people who were the ancient Greeks. The second book of the poem supplies, via a catalogue of the more than a thousand Greek ships that sailed to Troy, an account of the communities that in the mid-eighth century BCE regarded themselves as being united because they could enjoy poetry in Greek and because long ago their ancestors had fought together in the siege of Troy. The *Iliad* was performed at festivals where these self-governing Greeks, from diverse communities, met as equals in communal sacred spaces to worship their shared gods, and in doing so invented the competitive athletics contests (of which we possess a descendant in the modern Olympic Games) described in the *Iliad*, book 23. The poems recited at these gatherings were the collective cultural property of the independent-minded Greek warrior peasants wherever they sailed, and were fundamental to the transmission of their values. They remained so until the end of pagan antiquity.

Greek epic poems originated in oral composition; that is, they were developed through the process of being memorized, repeated, supplemented, and adapted over decades and (at least parts of) centuries. But between 800 and 750 BCE, Greek culture changed forever. Some resourceful Greek speakers, probably traders, borrowed the signs used by the ingenious Phoenicians to represent consonantal sounds, added some extra ones to indicate vowels, and used these signs to write down in Greek the poems of their already canonical authors. In inscribing this collective work, no doubt the poet-scribes (perhaps one was an individual really called Homer) made changes that ornamented the language and improved poetic structure.

I

The classical Greeks believed that the *Iliad* was aesthetically superior to other epic poems because it is not made up of episodes loosely strung together: it is instead unified by one incident during the Trojan War, a period of a few weeks when the great warrior Achilles became incandescently angry with both his overlord Agamemnon, who had disrespected him, and his Trojan enemy, Hector, after Hector killed Achilles's beloved friend Patroclus. The poem looks backward and forward in time to engage the listener with the war's antecedents and consequences, but is centered on this singular period.

The *Iliad* created the very core of the Greek sense of self for at least twelve centuries after it was first transcribed. Along with the *Odyssey*, it formed the basis of the education of every Greek speaker in ancient Mediterranean and Black Sea societies from the seventh century BCE; as Hegel noted, "Homer is that element in which the Greek world lived, as a human lives in the air."[1] Even from pre-Roman days it was not only the Greeks who studied this poem. For a thousand years, countless schoolboys living under the Macedonian or Roman empires, whose first languages were Syrian, Scythian, Nubian, or Gallic, learned their alphabet through the first letters of Homeric heroes' names, developed their handwriting by copying out Homeric verses, and honed their art of précis by summarizing individual books.[2] They also committed swaths of Homeric hexameters to memory; in Xenophon's *Symposium* 3.5, the Athenian Niceratus reports that his upper-class father required him to learn *all* of Homer by heart. They studied the literature in early manhood when they were learning to be statesmen, soldiers, lawyers, historians, philosophers, biographers, poets, dramatists, novelists, painters, or sculptors.

The *Iliad* continued to be read across the Byzantine world for another millennium. Its preservation, despite the rising Ottoman threat to Byzantine Greek culture, was guaranteed once the fourteenth-century humanist Francesco Petrarch acquired a copy via a contact in Constantinople. The text was translated into Latin, and began to be read in learned circles in the West; it was printed in the original Greek in Florence in 1488.

This precious printed edition unleashed a flood of translations into Latin and modern languages. It inspired painters, dramatists, and poets alike. Homer became central to the Western curriculum; European colonialism ensured that the *Iliad* made its way across empires on every continent. It was included, for example, in the list of Christian books supplemented by pagan authors constituting the *Ratio atque Institutio Studiorum*. Designed by Jesuits in Rome in 1599, this curricular guide and rulebook was exported across the planet by missionaries for the Society of Jesus.[3]

The curriculum was in turn adopted by Western humanists. In 1869, John Ruskin stressed that it does not even matter whether or not Homer is actually read, since "All Greek gentlemen were educated under Homer. All Roman gentlemen, by Greek literature. All Italian, and French, and English gentlemen, by Roman literature, and by its principles."[4] By the twentieth and twenty-first centuries, Homer had long since ceased to belong to the Western world, and had instead become a cultural property familiar on every continent.[5] The subterranean impact of the *Iliad* on our species' global psyche may not be overestimated. No later author could ever again make a fresh start when shaping a narrative or a visual representation of a quarrel between a self-regarding monarch and his able lieutenant, a council of gods, a siege war, an athletics contest, a viewing of an army from a city wall, a husband parting with his wife and baby, or a redemptive meeting of deadly enemies. The same goes for the poem's visions of production and consumption of materials: a hero's funeral, a smith at work, animals being sacrificed, workers reaping, trees being chopped down, or a vast ransom of precious metals put on public display. In literary critics' definitions of sublime art, especially "epic" poetry, massive scale and the evocation of the infinitude of natural resources became aesthetic requirements largely as a result of the tonal effects of the *Iliad,* which inspired and legitimized the activities of every agent of extractive industrialization and colonialism in history. The Homeric *Iliad* is therefore a foundational text in the culture not only of the Mediterranean world and Europe, but of the entire planet.

This epic poem also has the potential to be foundational to our struggle to save our planet from disaster, if we use it to expose the deepest contradictions underlying the environmental crisis that we humans have created. A priceless document of the mindset of the early Anthropocene, the *Iliad* shows how humans view their environments in ways that are informed by representations of nature in their art and literature, especially in canonical texts that have been widely translated, adapted, visualized, enacted, and included on the curriculum. How the poets of the *Iliad* depicted relationships between people and the physical world around them has fundamentally affected how we imagine those relationships, too.

Canonical artworks shape the way we see the world and act on and within it: as Gillian Rudd explains, our "response to the physical world is mediated by our social and literary creation of it."[6] Amitav Ghosh has proposed in *The Great Derangement* that the generic expectations of the Western novel, in which weighty individual characters act autonomously in front of circumscribed backdrops, and often struggle valiantly with short-ages of natural resources, have scarcely been congenial to evolving a more sustainable attitude to the natural world.[7] Perhaps the violence done to the environment over the past centuries has been authorized, if not exac-erbated, by the celebration of humans' exploitation of nature in the foun-dational *Iliad*.

Just over a century ago, the words "apocalypse" and "apocalyptic" ac-quired a new, secular meaning, relating to "a disaster resulting in drastic, irreversible damage to human society or the environment, esp. on a global scale; a cataclysm."[8] In the twenty-first century, apocalyptic fiction about the end times, and postapocalyptic fiction about the remnants of the human race that may conceivably survive, have become recognized subgenres, replete with doomsday fantasies.[9] Graphic narratives of epidemics, cata-clysms, permanent war, mass starvation, and extinction have gone viral, to the extent that the label "apocalypse porn" has appeared in popular cul-ture.[10] There is a danger here, especially the risk that real perils are being peddled as distracting voyeuristic or sensationalist entertainment while we fail to educate consumers into preemptive action.[11] Or, as Christopher Abram puts it: "By making such a catastrophe part of a general, inevitable schema of cosmological history, it may become less of a specific threat to one's psyche in the here and now."[12] But awareness of this mass psycholog-ical development can also help in analyzing—and perhaps encouraging—more socially responsible and aesthetically ambitious cultural phenomena.

Even more strikingly associated with the twenty-first century than the word "apocalypse" is "Anthropocene"—a term that was coined in 2000, although as a concept it has existed for more than 150 years.[13] Humans, *anthrōpoi*, inaugurated a new (*kainos*) epoch in which their actions would become a principal motor for planetary change. Paul Crutzen and Eugene Stoermer proposed that the changes humans had wrought in the global environment were of such geological and ecological significance that the current term used to describe our epoch, the Holocene, beginning at the end of the last Ice Age, had outlived its usefulness.[14] As Christopher Schliep-hake has warned, however, the term Anthropocene is inherently anthro-pocentric. There are good and bad visions of the Anthropocene, but both use humankind as the yardstick against which to measure environmental health and well-being. Another problem is that the word has the power of a collective singular to conceal cultural, historical, political, and socioeco-nomic disparities and heterogeneity by imposing an illusory homogeneity.[15]

The idea of the "Anthropocene" has caught on in public discourse, even though there remains debate about when the era started, a problem that a committee of scientists called the Anthropocene Working Group, formed by the International Commission on Stratigraphy, was established in 2009 to explore.[16] Proposals range from the mid-eighteenth-century industrial revolution all the way back to the end of the Pleistocene epoch and the agricultural revolution in Mesopotamia at least ten thousand years ago, and in other world communities in the centuries that followed.[17] The adoption of a conscious regime of productive agriculture entailed a world-wide adaptive shift in which *Homo sapiens* sought to wield greater control over natural resources in manufacturing ceramics, domesticating plants and animals, and making advances in other technologies including the production of weaponry.[18] It is undeniable that urban systems of premodern Mesopotamia centralized vast natural resources, bringing numerous ecological traps that became integral to the long-term effects we experience today.[19] By about 3500 BCE, for example, there had developed an increasingly intricate system of labor division in Mesopotamian farming-based societies. The specialized labor that drove the development of workshops, smithies, bakeries, and the timber, mining, and building industries can be seen in the *Iliad*, especially its craft similes and the account, for example, of Paris's beautiful private mansion (6.314–17),

> which he had constructed with the men who were at that time
> the best skilled builders in deep-soiled Troy;
> they had made him a chamber and a hall and a courtyard
> near Priam and Hector's on the highest part of the city.

Farming-based communities were ruled over by elite families like Agamemnon's and Priam's, "the masters of violence, who converted a comparative advantage in killing to control over politics."[20] It was at this time, proposed Ian Morris in his book *Foragers, Farmers and Fossil Fuels,* that daily energy capture rose above about ten thousand kilocalories per capita and the population of towns began to exceed ten thousand.[21] Extraction of natural resources to maximize the creation and accumulation of goods and resources, and population growth through urbanization, put humans on a maladaptive path from which we have not diverged, despite more recent environmental activism.[22] Let us hope that Habermas got it wrong when he argued that human sociocultural evolution has required "not-learning," a fundamental inability to learn from experience.[23]

Although there are legitimate fears that dating the onset of the Anthropocene too early might risk normalizing global environmental change, it has also been argued that the important point is that humans are the first geological agents to be able to reflect on their own agency.[24] I will argue that there are multiple clear signs of such reflection on anthropogenic landscape change in the *Iliad*. Moreover, countering Bronislaw Szerszynski, who claims that "the Anthropocene in all its geohistorical specificity" can really be said to have started only when humans became aware both of their agency in affecting landscapes and "in shaping climate," there are passages in the *Iliad* (admittedly only few) where human involvement in causing meteorological events is at least unconsciously implied.[25] Indeed, proposals for when the Anthropocene began include the period from which the *Iliad* emerged, when unprecedentedly intensive cattle farming and deforestation for shipbuilding took place.[26]

Demographers estimate that ten thousand years ago, before plant domestication, the total global human population was around six million. But by 1200 BCE it had grown to one hundred million, all felling trees to keep warm, cook, build, sail, farm, smelt, and fight.[27] Population growth in the Peloponnese accelerated exponentially in the thirteenth century. As the economy expanded, so did the population in the late Bronze Age, especially the thirteenth century. In the Plain of Argos, Mycenae, Tiryns, and Argos all expanded and new settlements were created. The number of villages in Messenia tripled, and their size doubled.[28]

Although the occasional polemical skeptic denies that there was substantial or consequential deforestation in ancient Greece, just as some question whether humans are responsible for ecocide and environmental collapse, there is an increasing volume of countervailing evidence that timber became a much rarer and more precious resource.[29] David Attenborough has described historical deforestation as striking "the crucial blow" to the ecosystem of the Mediterranean region.[30] The Homeric epic, although by many centuries the younger sibling of the Mesopotamian epics, is often felt somehow to represent the earliest file in the archive of human destructiveness, even to be the text that inaugurated modern consciousness and the breakdown of the prehistoric "bicameral mind."[31]

In the introduction to their edited collection *A Global History of Literature and the Environment*, John Parham and Louise Westling propose that potential ecocritical responses to literature fall into one of nine categories, most of which do not describe what I am doing in this book.[32] These responses, according to Parham and Westling, can be "reality"-based, and use literature to identify evidence of real-world problems such as drought

and deforestation over the centuries. This is of course possible, but, beyond emphasizing that the audience of the *Iliad* may have felt themselves
to be living in a post-apocalyptic society that remained vulnerable to arbitrary natural events, this is not my objective here. Nor do I investigate
celebrations of nature's beauty, since these are few and far between in the
Iliad; my goal is not directly to encourage others to make new literature,
films, or images that intervene in the environmental crisis, nor to trace
how the poets of the *Iliad* were trying, as Parham and Westling put it, to
shape "into coherence the scientific, social and epistemological complexi-
ties of their own and their culture's engagements with the environment"—
since with the rarest of possible exceptions they were certainly not doing
this consciously.

Instead, the category of criticism Parham and Westling identify that
has most informed my own methods is one in which researchers excavate
"a counter-history of half-buried or obscure literary traditions . . . that
describe eco-social interrelationships of humans with the world around
them."[33] This might indeed take the form of asking "how literature has
dramatized the permeable boundaries among animals, plant species, and
humans," or discerning dichotomies between "ecological respect for other
species and a social ecological pragmatism that regards nature, within limits, as a human resource."[34] Occasionally, my discussion will intersect with
postcolonial readings that address abuse of colonized environments, much
as Derek Walcott's Homer-inspired epic poem *Omeros* uses the hacking
down of the sentient trees of St. Lucia to evoke the violence of European
colonial exploitation of both Africa and the Caribbean.[35] Rather more often,
it engages with the political and class formations that underlie environmental impacts, like those explored in the work of Raymond Williams on
English pastoralism and its relationship with the industrial revolution.[36]

In order to show how profoundly the *Iliad* informs our imaginings of
armed conflict, one strand in my argument uses more recent poetry, fiction, and occasionally cinema and visual art to explore how interpretations
of ancient artifacts have an influence far beyond the academy. In doing so,
I pay homage to one of the most influential interpretations of the *Iliad* in
modern times, Simone Weil's extraordinary essay "*L'Iliade* ou le poème de
la force" (The *Iliad*, or the poem of force), which was written as war broke
out in 1939 and first published in 1940. I am in complete agreement with
Weil's identification of physical compulsion and violence, and their tragic
consequences, as the central subject of the epic poem.

One reason why Weil's essay in hindsight seems so significant is that

her account of the annihilation of Troy seems almost eerily to have anticipated the imminent genocides, as well as the wholesale destruction of entire cities by both conventional and nuclear bombs—events that the humans waging World War II were about to inflict on themselves and on the
other organisms with which they share the Earth. But another reason for
her essay's importance is that it theorized and came to represent the new
revulsion that emerged at the dawn of the twentieth century against the
Iliad's celebration of martial violence. It was the carnage of the Boer War
and especially bloodshed in the Western Front trenches that led to the
warriors of the *Iliad* being reassessed and sometimes dislodged from their
plinths as exemplars of manly heroism. More recently, too, the rise of feminism in both scholarship and public culture has led the epic's women,
locked in the brutal patriarchal system of commodity exchange it depicts,
to receive the gender-sensitive readings they deserved.

The urgency of the global ecological crisis must impel us to reassess
the Homeric warriors yet again, in terms of their rapacity toward their
natural environment. The *Iliad* can help us fight back against the current
dire new stage in the history of (in)humanity. A green reading will show
that the seeds of environmental catastrophe were already sown by warfare
millennia ago: on many occasions, as Brooke Holmes writes, "the antino-
mies that structure the modern sense of nature (nature and culture, nature
and art) seem easily traced to Greek origin."[37] It is time to reread this foundational text in the culture of the world in a new light, not only to prevent
World War III, but also to rescue our planet—and all the living organisms
we share it with—from disaster.

CHAPTER ONE

Changing Interpretations
of the *Iliad*

Sing, goddess, of the dreadful wrath of Peleus' son Achilles.

It afflicted the Achaeans with manifold causes of grief,

and sent to Hades the brave souls of many heroes,

leaving their bodies as spoils for dogs

and every bird of the air. Zeus' plan was being fulfilled.

So OPENS THE *ILIAD*, composed around 2,750 years ago. An erudite ancient scholar has commented on the last phrase, "Zeus' plan was being fulfilled," claiming that Earth begged Zeus to relieve her of the weight of the
multitude of people, who were behaving impiously.[1] So Zeus first brought
about the Theban War, which destroyed large numbers, and afterward the
Trojan one, "with Momus as his adviser, this being what Homer calls the
plan of Zeus, seeing that he was capable of destroying everyone with thun-
derbolts or floods." Momus, "Blame," was the sinister son of Night and
brother of Misery; he recommended, as an alternative to thunder or floods,
the Judgment of Paris. Thus the Trojan War came about, resulting in "the
lightening of the earth as many were killed."[2]

The scholar then quotes a fragment of the lost epic *Cypria*, which narrated the events preceding the war itself:

> There was a time when the countless tribes of humans roaming constantly over the land were weighing down the deep-breasted earth's expanse. Zeus took pity when he saw it, and in his complex mind he resolved to relieve the all-nurturing earth of mankind's weight by fanning the great conflict of the Trojan War, to void the burden through death. So the warriors at Troy kept being killed, and Zeus' plan was being fulfilled.[3]

A similar tradition was recounted in the early epic *Catalogue of Women* attributed to Homer's approximate coeval, Hesiod: "high-thundering Zeus was devising wondrous deeds then, to stir up trouble on the boundless earth; for he was already eager to annihilate most of the race of speech-endowed human beings . . . Hence he established for immortals, and for mortal human beings, difficult warfare . . . pain upon pain."[4] John Perlin suggests that these traditions are mythical responses to the historical depopulation as the Mycenaean world disintegrated.[5] The fall of Troy and the notion of an apocalyptic threat to the survival of the human race have thus been linked in the mythical imagination since the archaic age.

Classics and Apocalypse: "Lone and Level Sands"

Some myths are now being read allegorically in the growing field of geo-mythology, the study of how geology and myth are interrelated, in cases where myth partially conceals memories of real and catastrophic events, both natural and anthropogenic. Phaethon's fall from his chariot has been connected with actual comets that struck the planet; rather more plausible, sadly, is the argument that the pygmies mentioned, for example, in *Iliad* 3.1–7, represent literary memories of real-world small hominids living on the peripheries of the ancient Greek experience.[6] These tiny hominids had long been wiped out by the time the Homeric epics were finalized in approximately the form we have them today, by poets who, for convenience, are normally referred to in this book as "Homer."[7]

The somewhat obscure demonic entities known as the Telchines, who were already mentioned in the fifth century BCE by Xenomedes of Ceos, possessed such preternatural skill in metallurgy that they were associated with enchantment and wizardry.[8] They also had meteorological and environmental powers: in particular, they produced a toxin made from Stygian waters and sulphur that rendered all living organisms infertile and forced humans into wandering over the face of the Earth in search of food. It is probable that this paradoxical pairing of portfolios expresses an awareness of the terrible industrial pollution caused by smelting of copper, especially in the form of arsenical bronze.[9] Although the sources are few and abstruse, it is clear that the Telchines were destroyed by the gods' implementation of a natural disaster—a flood, Zeus's thunder, Poseidon (implying an earthquake), or Apollo in the shape of a wolf—as punishment for this environmental damage.[10]

The classical world has given us clear and now familiar images of societal collapse and apocalypse, both mythical and historical.[11] Hesiod believed that we were already living in the Iron Age, the worst age, when children dishonor their parents, siblings fight, and the social contract is forgotten. In this age humans are motivated entirely by self-interest, tell lies as truth, and are continually violent; in time, the gods will abandon mankind and we will destroy ourselves. Ovid and the mythographers, too, give us the story of the flood in the time of the *Ur*-ancestor Deucalion, a disaster inflicted on humans as punishment for their crime of human sacrifice. Zeus unleashed a deluge: the rivers became engorged, the seas rose, and only Deucalion and Pyrrha survived to perpetuate the human race.[12]

In his *Timaeus* and *Critias*, Plato gave us the image of Atlantis, a brilliant seafaring island civilization that became decadent and prone to overconsumption. It was annihilated by a terrible flood after earthquakes, and disappeared forever beneath the waves beyond the Pillars of Hercules. Scholars have long suspected that "folk memories" of tsunamis that beset Minoan culture on Crete underlie some at least of Plato's images of a submerged civilization in his *Critias*.[13] Plato also contrasts the Attica of his own day with that of nine thousand years earlier, before deforestation and consequent soil erosion had changed the coastline and made the landscape barren, compromising water supplies and wrecking the habitats of animals.[14]

The *Iliad* and Sophocles's *Oedipus Tyrannus* both open with communities in despair as their members die from plague, a theme that resonated during the protracted global experience of COVID-19. But in his *History of the Peloponnesian War*, Thucydides contributed the most famous plague narrative of all time, recounting the mass deaths of the Athenians and the concomitant anarchy that blighted their entire city-state in the early 420s BCE. He prefaces his most definitively scientific, Hippocratic-style passage, which analyzes the symptoms of the Athenian plague entirely in the third person, with a general description of the onset of the disease at Athens, and the unelaborated statement, "I had the disease myself and saw others suffering from it." The scientific description includes an account of the agony that the plague inflicted, and the small chance of survival it offered, along with the information that it could cause lasting damage to extremities, and even permanent blindness.[15]

Roman historians have supplied us with the terrifying account of the great fire of Rome, which raged for nine days in July 64 CE and destroyed two-thirds of the city's buildings.[16] Pliny's narrative and archaeologists have given us Pompeii and Herculaneum, obliterated by the eruption of Vesuvius fifteen years after the Roman fire, in 79 CE. Within years of Herculaneum's exhumation in 1738, by workers digging trenches for the foundations of a palace for Charles Bourbon, King of Naples, visualizations of the apocalyptic destruction in the Bay of Naples became the stuff of painting, fiction, and popular entertainment, a fascination that continues to this day.[17]

In the first century BCE, Diodorus of Sicily, following Hecataeus of Abdera, presented his readers with the Egyptian monarch Osumandias, whose colossal statue outside Egyptian Thebes, on the west bank at Luxor, bore the inscription: "I am Osymandyas, king of kings; if any would know how great I am, and where I lie, let him excel me in any of my works."[18] The poet Shelley transforms this reported inscription into his famous, dark exhortation against assuming that human ascendancy can last forever: "My name is Ozymandias, king of kings: / Look on my works, ye Mighty, and despair!"[19]

Shelley's poem, which meditates on collapses of nations and geopolitical empires, has recently inspired commentaries about the death of ideologies, the erasure of oppressed peoples from the historical record, and especially the impending death of organic life on planet Earth. These new Ozymandian deaths and disappearances involve interactions with classical authors other than Diodorus, especially the Greek tragedians and the poetic voice of the *Iliad*.[20] They have penetrated quintessentially modern cultural media and genres, from the multiple-prize-winning episode "Ozymandias" in the television crime-drama series *Breaking Bad*, to the lauded American comic book series *Watchmen* written by Alan Moore, drawn by Dave Gibbons and colored by John Higgins.[21] After appearing in monthly parts from 1986, *Watchmen* was published in a single volume in 1987. A runaway success not only financially but even among critics, it forever changed perceptions of the potential of comic-book formats when it made *Time*'s list of the hundred best English-language novels published since

1923. In due course it was made into a movie and an HBO television series as well.

In *Breaking Bad* and *Watchmen*, Ozymandias is implicated in a millennial nexus binding fears of moral decay, transnational capitalism, and environmental catastrophe. This new configuration of Ozymandias is a product of the nuclear age when our visions of global destruction stopped including the possibility of a few people surviving, and some continuity in organic life. As Maria Manuel Lisboa has put it, it was only when the enactment of annihilation "fell under the control not of God or nature but of human agency that it acquired cosmic possibilities. Robert Oppenheimer, gazing at the mushroom cloud which he had helped to create, understood this and acknowledged it not triumphantly but with post-lapsarian guilt" (in the famous statement "Now I am become death, the destroyer of worlds").[22] In that moment, he and the other perpetrators of the Manhattan Project "knew that for the first time in the history of the planet they really could put words into practice and reduce the planet to 'lone and level sands.'"[23]

Some have argued that Shelley's sonnet contains an optimistic statement about literary art's ability to transcend time and survive the fall of empires: the statue's frown and sneer, "Tell that its sculptor well those passions read/Which yet survive—stamp'd on these lifeless things."[24] One of Shelley's fragmentary poems, "Rome and Nature," written in 1819, consists of just three lines:

> Rome has fallen, ye see it lying
> Heaped in undistinguished ruin:
> Nature is alone undying.[25]

Yet we are at the chilling point in history when the fantasy that nature is undying, at least terrestrial nature, seems as vain as Ozymandias's faith three millennia ago in the eternity of his reputation. "Ozymandias" the poem will not outlive *Homo sapiens* any more than will the texts of Hesiod, Ovid, Plato, Thucydides, Pliny, Diodorus, the Greek tragedians and the *Iliad*, for that matter; nor, unless we act now, will it—or we humans—outlast the natural world on planet Earth.

The Limitless World of the *Iliad*

Although there are similar, if shorter, scenes in Mesopotamian literature, perhaps the most horrifying picture of planetary destruction in the entire ancient repertoire is to be found in the Greek *Iliad*. It is formed in the visual imagination, suitably enough, of the lord of the dead, Hades or Aidoneus, the "One Who Makes Things Unseeable." When the gods marshal themselves for war toward the climax of the epic, we are presented with a terrifying picture of a world split in two by an earthquake (20.56–65):

> Then the father of men and gods thundered terribly
> on high. From underneath, Poseidon shook
> the boundless earth and the steep peaks of the mountains.
> The roots of Ida with its many fountains were all shaken,
> and her summits, and the Trojans' city and the ships of the Achaeans.
> Underneath, Aidoneus, lord of those below, was terrified,
> and in his terror leapt from his throne and shouted,
> fearing that above him Poseidon the Earthshaker would cleave the
>   earth,
> and reveal his habitations to mortals and immortals,
> dreadful in appearance and slimy, so even the gods abhor them.

Hades feels the tremors shaking the very matter out of which the Earth above and around him is made. He fears that a vertical chasm will split the horizontal surface of the Earth to reveal his damp demesne in the deepest Underworld. Such an image could only be formed by a culture with some experience of the sensations caused by earthquakes and their landscape-altering consequences.[26]

A distinctive feature of Hades's view of the world is that its constituents are "boundless"—Earth is "without limit."[27] A crucial difference between our modern perception of the Earth and that of Homer's audience is that we now know there are limits to what the Earth can provide for humankind. The notion that we need to acknowledge the terrifying limitedness of natural resources was at last popularized in 1953 in Fairfield Osborn's *The Limits of the Earth*, where he observed that the history of Greece and Rome "assumes the character of a prologue to modern times."[28] It is impossible to imagine ourselves into a mindset where there were always new lands to conquer, new forests to chop down, and new seams of ore to mine. But we can begin to glimpse what it felt like by examining the idiom of infinitude that informs not only Hades's view of the limitless Earth, but also numerous other magniloquent Homeric expressions.

Timber from the forests of Ida is repeatedly said to be of unutterable extent, unspeakable, infinite (*aspetos*), as unspeakable (*thespesios*) as the bronze war equipment of the Achaeans when they march forth (2.457), its gleam reaching the heavens.[29] The flocks of sheep and goats that Iphidamas had promised as an additional bride-price for his wife before he left for Troy were unutterable (*aspeta*, 11.245), in addition to the more prosaic quantity of a hundred cattle he had already put down as a deposit. The Hellespont is "boundless," without limits (*apeiron*, 24.545), as is the land of Troy that raises its voice to lament Hector (24.776). The Trojans march making a clamor like cranes fleeing from wintry storms and "boundless (*athesphaton*) rain" (3.4). A false concept of ecological and environmental limitlessness is therefore as key to the depiction of the wrath of Achilles in the *Iliad* as is the never-ending questioning of the exact power relations between man and natural phenomena, man and god, and god and natural phenomena.

Priam tells Helen he once saw the "multitudes" of Phrygians encamped along the river Sangarius (now called the Sakarya, in modern-day Turkey, and a river in which pollution levels are currently rising at an alarming rate).[30] Andromache says that her father had received "ransom past counting" (*apereis'*) for her mother after taking her captive (6.427), a formula that occurs nine other times in the *Iliad*; Agamemnon applies it to the recompense he is prepared to pay Achilles (9.120).[31] Chryses brought "ransom past counting" for his daughter Chryseis (1.372); Peisander and Hippolochus say that their father will offer Agamemnon "ransom past counting" for their lives (11.134). Menesthios's mother was purchased by her husband in exchange for "a bride-price beyond counting" (16.178).

Another term that often implies an infinite quantity is the adjective *murios*, "myriad," which in the plural can also mean "ten thousand." The opening sentence of the *Iliad* claims that Achilles's wrath would inflict "measureless pains" on the Achaeans (1.2), but just before the end of the poem it is Priam who laments Troy's "countless (*muria*) sorrows" (24.639). Achilles predicts his death will cause Thetis "infinite (*murion*) grief in her heart" (18.88): Aeneas's fear of Achilles causes him "measureless grief" (20.282), just as terror of the enemy can be felt "infinitely" (*aspeton*, 17.332). The River Scamander intends to conceal Achilles's corpse beneath "infinite (*murion*) shingle" (21.319–20). And Hector had paid for Andromache with "innumerable (*muria*) bride-gifts" (22.472).

Homer sometimes provides visual images to help his listener envisage uncountable multitudes. Achilles says that he would not accept gifts from Agamemnon even if they were equivalent to all the wealth of Orchomenus, or Egyptian Thebes, "where the houses contain the most treasures, and there are a hundred gates, through each of which drive out two hundred warriors with horses and chariots"; not even gifts "as innumerable as grains of sand or dust" will suffice to reconcile them.[32] The Trojan forces are as myriad as leaves and flowers in spring (2.468); Iris, disguised as Polites, tells Priam that the Achaeans' numbers are as many as leaves or grains of sand (2.800). Hector addresses his "tribes of uncounted allies."[33] And the Achaeans' helmets and weaponry glitter as they flow thick and fast from the ships, like snowflakes sent fluttering by Zeus as they are blown along by the North Wind (19.357–61).

There is also outrageous exaggeration. Warriors are huge. After slaying Ereuthalion, the biggest and strongest man he ever saw, Nestor describes his foe's huge bulk sprawled prostrate over a vast area (7.155–56). Warriors claim to have achieved impossible feats of valor. Nestor later claims, implausibly, that in a long-ago battle against the Epeians he single-handedly felled fifty chariots and killed the two warriors riding in each of them: a solo display of valor and battlefield slaughter (*aristeia*) of no fewer than a hundred enemies in one incident (11.747–49). The number of items that warriors possess is always overstated, too. The Trojan king Erichthonius supposedly had three thousand mares grazing in his pasture-lands (20.221).

A special kind of enormity is reserved for descriptions of the outsized world inhabited by the gods. The gods themselves are huge. Athena fells Ares and he stretches out across seven plethra: a single plethron is approximately one hundred square feet, or a quarter of an acre (21.407). The sense of unbelievable scale even enters the poem's acoustics. Ares and Poseidon both bellow as loud as nine or ten thousand warriors in battle (5.859–60, 13.148–49; see also 18.219–20). Hera's chariot has curving bronze wheels with eight spokes, whereas nearly all Bronze Age and Early Iron Age depictions have just four, and only occasionally six.[34] Athena's helmet has two horns and four golden bosses, and is "fitted out with the men-at-arms of a hundred cities," an image designed to suggest the huge size of both helmet and wearer.[35] It is beyond the capacity of the human imagination to visualize a helmet on a scale that can accommodate individual depictions of a hundred cities and their attendant soldiers, presumably in multiples of a hundred.

But infinity is temporal as well as spatial, quantitative, and sensory. One aspect of how the *Iliad* portrays that all things are infinite in duration, size, quantity, and effect on the senses is difficult to explain without referring to the sound of the ancient Greek language. Infinity is often indicated by adjectives with a prefixed "privative alpha," an "*a*" with a negativizing sense (in English the equivalents are the prefixes "un" or "in" or the suffix

"-less"). The effect of the "*a*" sound is sad and solemn, like a sigh, and its repeated use conditions the poem's mournful emotional and acoustic impact. A reader could try saying the transliterated Greek words here one after the other to feel the effect: the emphasis normally goes on the alpha or the syllable after it. Agamemnon's scepter, made by Hephaestus, is "forever imperishable" (*aphthiton*, 2.46). Helen's grief is unceasing (*akrita*, 3.412), as the tales told by old men can be, says Iris in disguise as Polites to Priam. War is unabating (*aliastos*), as can be battle, din, and lamentation (14.57, 12.471, 24.76). The fire that gleams from Diomedes's arms is unwearying (*akamaton*, 5.4), an Ancient Near Eastern motif, too.[36] The two Ajaxes, Achilles, and the Argives are "insatiable (*akorētoi*, 12.335, 20.2, 14.479) of war." Both divine laughter and human shouting can be "inextinguishable" (*asbestos*, 1.599; 11.50). The pain the Trojans feel when Sarpedon dies is "unbearable" (*ascheton*, 16.548–49). A fit of trembling can be boundless, or unceasing (*aspeton*, 17.332). The sense of infinitude affects everything. Hera "rages unceasingly" (*asperches*, 2.32); Achilles believed his wrath against Agamemnon would never end (16.61); Achilles tells his horses he intends to drive the Trojans to a "surfeit of war" (19.423).

Unlike the *Iliad*'s humans, its immortals, at least the supreme couple, Zeus and Hera, and their favorite messenger, Iris, do seem aware that there are specific (if extremely remote) limits inherent in the cosmos, at least to the Earth and sea, and that a people known as the Ethiopians live far away near the streams of Ocean that encircle the world (23.205–6). But other geophysical boundaries lay deep beneath the Earth in Tartaros, where neither sun nor wind can reach them, and thus where they remain unknowable by living human beings. Zeus tells the furious Hera that he is unconcerned about her anger, even if she should go to the deep place where Iapetus and Cronus now reside at the "nethermost bounds (*peirath'*) of earth and sea" (8.478–79). Hera lies to Aphrodite, saying that she is about to travel to "the limits (*peirata*) of the all-nurturing Earth" (14.200), where Ocean, "from whom the gods are sprung" (14.301) and Tethys live, but are endlessly quarreling.[37] Iris tells the Winds that she needs to travel via the stream of Ocean to the land of the Ethiopians (23.205–6). We are required to imagine Hera grasping the entire bounteous Earth in one hand and the shimmering sea in her other when Hypnos (Sleep) prescribes how she is to take her oath to him (14.271–73).

This distinctive idiom of gargantuan scale and unboundedness, "Homer's characteristic evocation of dimension beyond measurement," is a constituent of the poem's grandeur imitated by emulators and parodists such as

Aristophanes and Lucian, and admired by ancient literary critics.[38] The literary critic known as Longinus regarded it as lending Homeric epic, especially the *Iliad*, true sublimity, or elevation. Longinus identifies as sublime the evocation and deliberate magnification of huge distance, between Earth and heaven, encompassed by Eris's stature or the length of divine horses' strides (4.441–43; 5.770). In Longinus's conflation of two passages about Poseidon, whose coming makes forests, mountains, Troy, and the Achaeans' ships all quake, the literary critic claims that Homer singles out a "majesty" that surpasses even the Theomachy.[39] Homer is himself "swept away by a whirlwind" when he describes Hector raging like Ares, wielder of the spear, like a wildfire among the mountains in the thickets of a deep wood (16.605).[40] Longinus praises Euripides's intermittent grandeur by quoting a simile from the *Iliad* in which Achilles is compared with a wounded lion working himself up to fight; both the critic and the tragedian were responding to the unforgettable imprint on the poem that Homer's refashioning of the lion has left, whether as threatening marauder or as victim of the human hunt.[41]

The *Iliad*'s evocations of scale, infinity, and the chaotic beauty of elemental and feral nature are some of the very characteristics that make it speak so loudly to a modern age riven with anxiety about Armageddon. During Achilles's apocalyptic fight against the River Scamander, Homer introduces a crucial simile that encapsulates the conflicted relations between man and environment that characterize the entire world of the poem. The great river god behaves like a stream of water whose course a gardener has tried to divert (21.258–64):

> It was as when a man guides the flow of a stream of water from a
>   murky spring, leading it
> through his plants and gardens with a mattock in his hands, creating
>   dams in its course.
> As it flows along, all the pebbles underneath are swept along with it,
> and it rolls quickly onwards with a gushing sound
> and it overtakes even the man who is guiding it.
> That was how the streaming wave continuously overtook Achilles,
> despite his swiftness. For the gods are more powerful than men.

Human beings know how to interfere in nature in order to make it serve human ends, but we cannot predict the full consequences of that interference because something—the ancients called it the gods—is more powerful than we are. It was impossible for me not to be reminded of this simile while I was completing this book in September 2023, when the collapse of two dams in Derna, Libya, after a torrential storm caused the deaths of as many as twenty thousand people.[42]

Earlier Responses to the *Iliad*

Ever since the first printed texts of the Homeric epics appeared in 1488, the poems have been seen as culturally foundational. The meanings attached to the *Iliad* carry weight and authority commensurate with its foundational status. In one sense, this status is misleading. The poems contain material inherited from an Indo-European tradition shared with Sanskrit culture, and represent a late stage in the evolution of Ancient Near Eastern mythical narrative poetry, above all the Sumerian *Epic of Gilgamesh*, in societies that had reached peaks of sophistication millennia earlier.[43] Although it also contains material similar to the biblical story of Noah and the flood, *Gilgamesh* shares elements with both the *Iliad* and the *Odyssey:* like Achilles, Gilgamesh has a beloved comrade-in-arms whose death he can scarcely abide and with whose ghost he holds a painful conversation.[44]

Both Gilgamesh and Achilles are sons of immortal mothers who interact and discuss the future with them and human fathers who stay away from the main action. Both heroes are big, beautiful, impulsive, proud, and emotional, and they wrestle with metaphysical questions, especially mortality.[45] It is not just the plot and characters that are similar; Johannes Haubold has shown how the Mesopotamian epic shares with the *Iliad* both poetic atmosphere and evocative tropes concerning the human condition.[46] Over the past three decades, the relatively late place occupied by the Homeric epics in the development of world literature has been proven. But the *Epic of Gilgamesh* was neither rediscovered nor deciphered until the second half of the nineteenth century.[47] The privileged place offered to the Homeric epics in Western and imperial curriculums for several centuries preceding the painful process of decolonization in the twentieth and twenty-first centuries has lent them an autonomous cultural presence of unusual persistence and potency.

There have, of course, been shifts in perceptions of both poems, especially the *Odyssey*, which, until recent times, was far more familiar to a wide cross-section of the world's population than was the *Iliad*.[48] With a few notable exceptions, until relatively recently, the heroes of the *Iliad* were celebrated somewhat simplistically as supreme exemplars of military valor.[49]

Ever since the Renaissance, the warriors of this great epic had been held up as ideals of authoritative manly prowess. In the newly independent United States, Thomas Jefferson repeatedly re-read the *Iliad* to immerse himself in what he saw as its profound wisdom and applicability to the issues facing his young republic.[50] The Achaeans' manly prowess, displayed on a foreign field in order to punish eastern barbarians' arrogance and cupidity, became increasingly valuable in an era celebrating imperialism and manly feats of derring-do against international rivals and colonial subjects.

To pick a typical example, Achilles was identified with Horatio Nelson in a panegyric recited to the Belfast Literary Society in 1806, *The Battle of Trafalgar, a Heroic Poem.* The author was the classically educated Protestant man of letters, Reverend William Hamilton Drummond. In the poem, Achilles's *aristeia* and Nelson's become indistinguishable:

> Next fierce Achilles, o'er the azure field,
> Lifts the broad splendour of th' immortal shield;
> Sheathed in refulgent panoply divine,
> He moves in flames along the glancing brine:
> So stern, so ruthless, so athirst for blood,
> He strode to battle through Scamander's flood;
> With direst rage, implacable he burns,
> And all his fury on Iberia turns;
> Shakes the keen spear before his buoyant car,
> And leads his British Myrmidons to war.[51]

The muscular Christians who subsequently ran Victorian Britain still saw themselves as plucky British Myrmidons and revered the Homeric warrior ethos.

William Gladstone himself, in *Studies on Homer and the Homeric Age,* wrote of "Homer's ever wakeful care in doing supreme honour to Achilles," and "the military grandeur of the hero," calling Achilles "the great warrior" and "the prime and foremost pattern of the whole Greek nation," and arguing that Homer took care to give warriors killed by Achilles names "in preference to letting Achilles slaughter a crowd of ignoble persons." He speaks of "the splendour of the fame of heroes," and how Homer "had lifted Achilles and Ulysses to a height surpassing" that of the others. Other heroes receive accolades, especially Sarpedon, who everywhere "plays his part with a faultless valour, a valour set off by his modesty, and by his keen sense of public duty according to the strictest meaning of the term."[52]

Achilles and Pallas Athena (Achilles Shouting from the Trenches), by Thomas Woolner. (Photograph by George P. Landow, via victorianweb.org)

Royal Academician Thomas Woolner was commissioned to design a marble triptych, a panel to ornament the plinth supporting a bust of Gladstone for the Bodleian Library in Oxford, which was presented to the subscribers in 1866. Woolner was the only sculptor among the original founding members of the Pre-Raphaelite Brotherhood. Between the two flanking scenes with Thetis is *Achilles Shouting from the Trenches* (the illustration here is a bronze copy), in which a muscular and naked Achilles, with his arm raised, is protected by Athena's shield, spear, and aegis as he bellows across the battlefield. This image was much admired by critic William Michael Rossetti (Dante Gabriel's brother), among others; Rossetti described the depiction of the "superhuman cry" as "a most vigorous and admirable composition."[53]

It was the American Civil War that seems to have turned the tide on readings of the *Iliad* toward a critique of its glorification of war and models of masculine heroism. Parallels were certainly drawn by combatants in that war, as we see in the 1947 book *The American Iliad: The Epic Story of the Civil War as Narrated by Eyewitnesses and Contemporaries* by Otto Eisenschiml and Ralph Newman. But the reappraisal of the militarism of the *Iliad* was most clearly articulated in the 1895 novel *The Red Badge of Courage* by Stephen Crane, which uses distorted parallels with the *Iliad* to satirize the battlefield performance of its protagonist, Henry Fleming, and the absurdity of war more widely.[54]

In Crane's literary allegory, the Union ranks are divided by dissension, much like the Achaeans were at the opening of the *Iliad*. Fleming's mother, far from being a goddess like Achilles's mother, Thetis, is a poor widow struggling on her farm. Fleming is motivated by rage at the commanders' disrespect of the common soldiers, but ultimately regrets this rage, which he feels has caused him to behave like an animal. He concludes that war is amoral; success in it is not a matter of courage but of random chance.[55] Or as N. E. Dunn writes, "Warfare is no longer a matter of heroic hand-to-hand combat, it is a matter of efficient mass murder."[56] The thunder and lightning released by Zeus in the *Iliad* is transformed by Crane into the blinding light and din of artillery fire and the insistent beating of military drums.

During the Boer War, the South African Conciliation Committee took up the unfashionable position that British imperialist conduct in South Africa was wrong. One of its supporters was Greek scholar Gilbert Murray, whose radical reassessment of the Trojan War was articulated by a notorious production at London's Court Theatre of his translation of Euripides's *Trojan Women*.[57] At around the same time, in eastern Europe, Lesia Ukrainka was rethinking the entire *Iliad* from a feminist perspective, denouncing the poor treatment of women and the working classes by imperialist male militarism, in her Ukrainian-language *Cassandra* (1908).[58] And by World War I, the *Iliad* itself was being invoked in poems by combatants who expressed not only their despair at the wholesale slaughter they were witnessing, but also their own fear of dying on the battlefield.[59]

One of these combatant poets was Patrick Shaw-Stewart, a Welsh-born Oxford graduate from a military family who had joined up in 1914 and served with Rupert Brooke. In a period of relative calm before combat began at Gallipoli, just across the Dardanelles from the site of ancient Troy, he wrote his single poem, "I saw a man this morning." In the penultimate stanza he asks Achilles directly how difficult was the death he experienced, before asking the hero to shout for him, alluding, in a much darker way, to the episode where Achilles shouts from the trenches—the one that Thomas Woolner had recreated in his sculpture:

> Was it so hard, Achilles,
> So very hard to die?
> Thou knewest and I know not—
> So much the happier I.

> I will go back this morning
> From Imbros over the sea;
> Stand in the trench, Achilles,
> Flame-capped, and shout for me.

Shaw-Stewart was indeed eventually killed in France in 1917.[60]

The *Iliad* was similarly subverted in David Jones's brilliant modernist prose poem *In Parenthesis*, finally published in 1937, eighteen years after he had been demobilized. There are similes reminiscent of Homer's grandest: sections that open with Iliadic sunrises, ritualized reiterative formulaic language, quasi-Homeric epithets, and "clear-voiced heralds."[61] The richest classical allusions are delayed until the section that is also the most formally "poetic," containing both abbreviated colometry and sustained rhetorical flow. This is the boast of Dai Greatcoat, Jones's homage to the "flyting" speech of Diomedes to Glaucus in *Iliad* 6.119–43, as well as an ironic salute to David Lloyd George.[62]

But the war that made it impossible to embrace the Achaean displays of violence in the *Iliad* with uncomplicated enthusiasm was World War II, and the author who was most instrumental in the sea change of interpretation was Simone Weil. This political activist, philosopher, and eventual convert to Christian mysticism was born in 1909 to Jewish parents in Paris. Originally a pacifist, she had tried unsuccessfully to fight for the Spanish Republicans; in 1939, as Germany invaded Poland and passed ever more stringent laws against Jews, she drew on her excellent knowledge of ancient Greek to reinterpret the *Iliad* as the Poem of Force. "The true hero, the true subject, the centre of the *Iliad* is force. Force employed by man, force that enslaves man, force before which man's flesh shrinks away."[63] *L'Iliade ou le poème de la force* was first published in 1940 in *Les Cahiers du Sud*; the first English translation was not published until 1945. It has since been frequently reprinted in many languages and is on the curriculum nearly everywhere the *Iliad* is taught, whether in ancient Greek or modern-language translation.

Less well-known, but increasingly read by classicists today since an important discussion of it by Seth Schein, is Rachel Bespaloff's 1943 *De l'Iliade*, first published in New York shortly after Bespaloff moved there as a Jewish refugee from occupied France. A philosophical expert in existentialism and phenomenology, Bespaloff explicitly compares the ancient and contemporary war-riven worlds, and finds the contemporary one to be lacking even the flickers of pity, humanity, self-respect, and dignity that were manifested, for example, in Achilles's eventual concession to Priam. She, like Weil, sees force as a central concern of the poem, but it can be positive, as in the case of Hector, "Ilion's protector, defender of a city, a wife, a child, . . . the guardian of the perishable joys."[64]

The substitution of Hector for Achilles as the true hero of the poem was clearly discernible even before Weil, in Wolfgang Schadewaldt's 1938 celebration of Hector's heroism as carrying a profound message for all human beings.[65] This work foreshadowed the general tendency in postwar responses to this day, both within and beyond the academy. In 1975 James Redfield gave his pioneering *Nature and Culture in the Iliad* the subtitle *The Tragedy of Hector*; indeed, Hector's selflessness and the pathos of his predicament dominate many studies published between Jasper Griffin's *Homer on Life and Death* (1983) and Lynn Kozak's feminist *Experiencing Hektor: Character in the Iliad* (2016). Outside academic classics, the tragedy of Hector's family is a core theme in Wim Wenders's 1987 movie, *Der Himmel über Berlin*. The elderly storyteller named Homer muses as he looks through the annals of history and the spaces on the globe: "My heroes are no longer the warriors and kings but the things of peace equal to one another." The problem is that "epic" media have not proved conducive to amity and concord: "no one has so far succeeded in singing an epic of peace."[66]

In Wenders's film, as we listen to "Homer's" thoughts on the difficulty of creating art out of the experience of peace, we are shown footage from the end of World War II. Soldiers supervise women trying to identify the corpses of their children after the bombing of Berlin. The camera lingers on a tiny cadaver, with eyes closed and mouth open, screaming noiselessly. The baby's lifeless body lies beneath a high wall scarred with bombardment.[67] Behind this shocking image lurks the original Homeric epic of warriors and kings, the *Iliad*, for in book 6 we meet the infant Astyanax, son of Hector and Andromache, on the wall of Troy—the wall from which he will be thrown to his death by the enemy.

In Tony Harrison's poem "A Cold Coming" (taking its title, with deliberate irony, from the first line of T. S. Eliot's "The Journey of the Magi"), published in *The Guardian* in 1991, Harrison makes the deceased "charred Iraqi" in his burnt-out vehicle assume that the Western press had avoided images of Gulf War Iraqi dead for the same reason that he doubts "victorious Greeks let Hector/join their feast as spoiling spectre."[68] A year later, in Harrison's extraordinary film-poem *The Gaze of the Gorgon*, the most important co-text is the *Iliad*. The opening frame consists of a quotation from Weil's *The Iliad, or the Poem of Force*. It reads: "To the same degree, those who use force and those who endure it are turned to stone."[69] In Harrison's film, the abuse of Hector's corpse is key. It is portrayed in a large painted fresco, Franz Match's *Triumph of Achilles* (1892), which is installed in the Achilleion, the summer palace built on Corfu in 1890 for Empress Elizabeth of Austria. The voiceover (the German Jewish poet Henrich Heine) says that Homer's "*Iliad*/was the steadiest gaze we've had/at war and suffering."[70]

The camera then surveys a statue of Achilles by Johannes Götz (1909), commissioned by the anti-Semitic Kaiser Wilhelm II: this Achilles is a terrifying, upright hoplite, with a gorgon carved on his shield and the inscription "The greatest German to the greatest Greek" on its plinth. The camera then returns to the painting, to linger not on Achilles riding his chariot but on the women of Troy, especially Andromache, fainting at the sight of her husband's cadaver.

> The soon-to-be-defeated rows
> of Trojans watch exultant foes
> who bring the city to the ground
> then leave it just a sandblown mound.

These verses correspond with specific passages in the *Iliad*—Andromache's crazed dash to the walls when Achilles has killed Hector (22.437–72), and the simile of the boy kicking down a sandcastle (15.362–64). The poet concludes the commentary on this painting with a return to the parallels between the Holocaust and the destruction of Troy, remembered only because of Homer:

> A whole culture vanished in the fire
> until redeemed by Homer's lyre.
> A lyre like Homer's could redeem
> Hector's skull's still-echoing scream.[71]

The final sequence consists of horrific images of charred Iraqi corpses and a burnt-out tank, all caused by Operation Desert Shield, with the Gorgon's face superimposed.[72]

These are not the first nor even the most shocking images in the film. The newsreels from the mass graves of World War I, and the concentration camps of World War II, are breathtakingly dreadful. But these follow some of the most memorable verses in the film:

> The *barbitos*, the ancient lyre,
> since the Kaiser's day,
> is restrung with barbed wire.
> Bards' hands bleed when they play
> the score that fits an era's scream,
> the blood, the suffering, the loss.
> The twentieth-century scream
> is played on barbed-wire *barbitos*.[73]

These lines are recited over a beautiful carved image of that particular type of large, heart-shaped, deep-toned ancient lyre.

Two years later, Michael Longley's poem "Cease Fire," a paraphrase and condensation into fourteen lines of *Iliad* 24.503–634, appeared in the *Irish Times* after the IRA had declared a ceasefire.[74] At the same time, Jonathan Shay's *Achilles in Vietnam* (1994) used the *Iliad* psychoanalytically. Shay examined how the symptoms of what is now known as post-traumatic stress disorder, and the military environment in which that type of mental illness develops, are adumbrated by the conduct and circumstances of the arguably psychotic Achilles: he is bereaved, humiliated, disillusioned with the moral point of the war, and let down by his commanding officer.[75] Such readings as these by Wenders, Harrison, Longley, and Shay made the *Iliad* a crucial text in increasing the revulsion against warfare that made its outbreak less likely, at least from the 1950s until the 1990s in Europe. But these writers all emphasize the toll taken by war on the human psyche as a discrete phenomenon rather than an integral part of the total natural world. The need for an extended new reading that incorporates environmental as well as human social and psychological damage has become pressingly urgent. As Jean Wahl put it in 1943, "In these times of hardships, it is natural that Western thought turn toward its origins."[76]

New Resonances of the *Iliad*

In a book published fifteen years ago, I tried to explain the magnitude of the cultural responses to the *Odyssey*, transculturally across the planet, in terms of its aesthetic variety and originality, its generic hybridity, its psychological and sociological resonances, and its intellectual profundity.[77] The impact of the *Odyssey* on European and subsequently world culture has quantitatively—and some would argue qualitatively—surpassed that of not only all other ancient epics, but also the *Iliad*. Yet in the twenty-first century there has been an unprecedented explosion of artworks in diverse media reconfiguring the more gloomy and martial Homeric poem. Filmmakers, poets, and novelists, notably women, have rewritten it, to make its contents familiar as never before, except perhaps in the eighteenth century, between 1715 and 1720, when the evergreen English translation by Alexander Pope was published serially.[78]

There are several different reasons for the *Iliad*'s recent renaissance. First are its historical and political background and the transnational themes of continents in combat, war crimes, captivity, and migration. We cannot ignore the influence of the global sociopolitical reordering in the wake of the attacks of September 11, 2001. The US invasion of Iraq exacerbated preexisting tensions between the United States and the Muslim world, even though Pakistan, Indonesia, the Persian Gulf monarchies, Jordan, Egypt, and Morocco officially remained American allies.[79] The "enemy" in George W. Bush's "War on Terror" was expanded to include all parts of all countries in which Al-Qaeda and the Taliban were believed to operate as well as the jihadi strain of Sunni nationalism. There were attempts to aid the overthrow of the Ba'athist regime in Syria and the Shiite ayatollahs in Iran.[80] After 9/11, the Bush administration had some success in bringing to power previously losing sides in Afghanistan and Iraq, but the violent forces of civil conflict that this unleashed have created dangerous instability rather than liberal democracies across the Middle East.

A few commercial fiction writers jumped on the Trojan War bandwagon in the wake of the 2004 movie *Troy*.[81] But the first internationally important *Iliad*-inspired novel of the twenty-first century was *The Songs of the Kings* (2002) by Barry Unsworth, winner of the Booker Prize. Alongside Aeschylus's *Oresteia* and Euripides's *Iphigenia in Aulis* (with *Iphigenia in Aulis* also becoming suddenly popular in the wake of 9/11), the novel's primary co-text is the *Iliad*, even though the events it portrays—the days leading up to the sacrifice of Iphigenia at Aulis—precede the action of the epic.[82] In the first half, the focus is on the creation of Homeric epic, the precise mechanisms whereby truth is transformed, through "spin," into a manufactured narrative that suits the interests of history's winners.[83]

Unsworth saw the Trojan legend as a series of tragic events created by "spin-doctors" (especially Odysseus); he was motivated by his horror at the rebranding of the Labour Party by the then British government, especially by Tony Blair's lieutenants Peter Mandelson and Alastair Campbell, and the specious case made for supporting the aggressive military policies by the United States post-9/11.[84] Unsworth's novel brushes the Trojan War story against the grain, lending voice to the ancient underclass. Unsworth was himself born into a working-class mining family in County Durham, England, and all his fiction demonstrates an acute sensitivity to the casual brutality of the language with which powerful people address the powerless. The date when this novel was published, a year after 9/11, certainly made the cynical, warmongering exploitation of the ignorant by Unsworth's obnoxious spin-doctor Odysseus seem terrifyingly topical.

It is impossible to quantify the precise instrumentality of global politics in Warner Brothers' decision to make the movie *Troy*, although there is no doubt that shortly after 9/11, in November 2001, the head of the Motion Picture Association in North America, Jack Valenti, addressed a meeting attended by the presidential adviser Karl Rove and executives from all the major media conglomerates.[85] He invited his audience to keep in mind the global scope of their industry's ideological influence: "We are not limited to domestic measures. The American entertainment industry has a unique capacity to reach audiences worldwide with important messages."[86]

Decades after Hollywood had stopped producing blockbuster movies set in ancient Greece, within a few years of 9/11 no fewer than three major star-studded films about ancient Greeks had been produced. Each focuses on one of the three major military offensives against Asiatic peoples that structure Greek history: *Troy*, which was directed by Wolfgang Petersen (2004), set in the late Bronze Age and loosely based on the *Iliad*; *Alexander*, directed by Oliver Stone (also 2004), which charts Alexander the Great's eastern expedition leading up to his death in 323 BCE, a narrative derived from ancient historians including Diodorus, Arrian, and Plutarch; and *300*, directed by Zack Snyder (2007), which screens the battle of Thermopylae during Xerxes's invasion of Greece in 480 BCE. The ultimate historical source for *300* and its portrayal of the battle of Thermopylae is Herodotus, *The Histories*, book 7, but the influence is distorted by the intermediary of the comic-strip illustrated novella version, created by the reaction-ary graphic artist Frank Miller in 1998.[87] Although the success of Ridley Scott's *Gladiator* (2000) had nudged the movie industry into reviving the tradition—so popular in the 1950s and 1960s—of creating spectacular movies set in classical antiquity, the emergence of an urgent new global narrative of a long war between West and East, with a dividing line falling somewhere in the eastern Mediterranean, was surely a factor in the gene-sis of all three "Greeks versus the East" films of 2004–2007.

Troy was written by David Benioff in 2002, soon after 9/11; a draft was circulating by November 2002.[88] The filming approximately coincided with the invasion of Iraq. The outstanding performance was Brian Cox as the ruthless, cynical Agamemnon, determined to use Helen's elopement as the pretext for a military campaign motivated by personal greed. The director was Wolfgang Petersen, who had studied Greek at school, and had made his name with *Das Boot* (1981), a film he wrote and directed that was set during another cataclysmic conflict, World War II. When reflecting on the launch of the Iraq invasion in March 2003, he said, "I couldn't believe it. I thought, it's as if nothing has changed in 3,000 years. People are still using deceit to engage in wars of vengeance." He recalled that the similarities between what happened at Troy and what was happening in Iraq became all too obvious: both Agamemnon and President George W. Bush used pretexts to conceal their true motives in waging what were essentially wars of conquest.[89]

The ongoing military conflicts, along with the stellar cast, spectacular cinematography, and refreshing commitment to telling a tragic story in a sincere, unironic way, helped the movie *Troy*—despite its failure on several scores as a reception of the *Iliad*—to achieve worldwide success. The film certainly lies behind the subsequent penetration of the *Iliad* to a level of popular culture previously reached by hardly any other classical text except the *Odyssey* and *Aesop's Fables*. A 2006 advertisement for Amstel Brewery presented itself as a lost scene from the movie in which some Trojans fail to drag the wooden horse through the gates and drink lager instead.[90] The *Iliad* was chosen for adaptation in 2007–2008, along with *The Jungle Book*, *The Last of the Mohicans*, *Moby-Dick*, and *The Three Musketeers*, as one of the earliest in the *Marvel Illustrated* imprint series of comic-book adaptations of literary classics.[91] The movie *Troy* is likely to have accentuated the taste for archaic "epic" content in Heavy Metal music, and certainly informed "Iliad," the fifth track on the debut album *Of Secrets and Lore* made by Italian band King Wraith in 2015.[92] In 2018, BBC/Netflix launched its televised miniseries *Troy: Fall of a City*. The choice of Troy for the second installment of the *Total War Saga* video game developed by Creative Assembly Sofia, released in 2020, is another recent example.[93]

Although a few fiction writers jumped on the Trojan War bandwagon in the wake of *Troy*, the first internationally important *Iliad* novel after Unsworth's, while undoubtedly connected with the War on Terror, seems to have little connection with the movie. The prizewinning *Ransom* (2009), by the Australian David Malouf, retells the *Iliad* from books 22 to 24—the story of Priam as he goes to Achilles to plead for the return of the body of Hector.[94] Although *Ransom* was not written until the War on Terror was well under way, and did not appear until 2009, Malouf had written a poem in 1970, "Episode from an Early War," that relates the image of the lacerated corpse of Hector to his own experience of growing up as a small child during World War II and the genocide of the Jews.[95]

Several aspects of *Ransom* directly echo Weil's analysis, especially the emphasis on the way that neglect or violence reduces people to the status of mere object, and the sense of apocalypse already implicit in her essay: "The whole of the *Iliad* lies under the shadow of the greatest calamity the human race can experience—the destruction of a city."[96] This apocalyptic vision of total annihilation of a community, which made the *Iliad* feel relevant to Weil at the beginning of World War II, also features in *Ransom*. Somax, the ordinary Trojan who drives Priam into the Greek camp, articulates a stark vision of what will become of his nation, "later—when Troy has become just another long, windswept hilltop, its towers reduced to rubble, its citizens scattered or carried off . . . into exile and slavery," language that evokes images of Baghdad's civic infrastructure smashed to pieces by the 2003 US-led air strikes.[97] Malouf's story ends with an agonizing flash-forward to the moment of Priam's violent death at the hands of Achilles's son, Neoptolemus, heralding the annihilation of Troy.

The invasion of Iraq also helped to prompt two important poems by English women poets who are trained classicists. The first, "Fresh Meat: A Perversion of Iliad 22," by Josephine Balmer, published in 2004, is a relatively short but eerily beautiful account of the climactic last fight between Achilles and Hector (22.25–360).[98] Balmer takes her cue from Hector's Homeric death speech, reading the encounter from the Trojan's perspective as a very physical meditation on the process of dying:

And I watched him as he scoured my skin for that one soft spot
where the flesh might best be pierced; as he found it on my neck

between jugular and wind-pipe, and then drove home the point,
leaving me just breath enough to beg for more.

Alice Oswald's *Memorial* followed in 2011. Both poets have responded to
a heartbreaking passage in which Priam, after foreseeing his own aged, ugly
cadaver being rent by his own guard dogs, articulates the epic's uncomfortable beautification of the dead bodies of young men (22.71–73):

> But when a young man is killed in battle
> and lies mangled by the sharp bronze, everything looks seemly:
> although he is dead, whatever part of him is seen appears beautiful.

Oswald focuses on the deaths of individual combatants during the Trojan
War; although no explicit connection is drawn, the appalling fatalities subsequent to 9/11 provide the background for all eighty-four pages of her
somber text. Between 2003 and the publication of *Memorial* in 2011, more
than 162,000 Iraqis lost their lives as a result of the conflict; in the same
period, nearly seven thousand American and several hundred British military personnel died in Iraq or Afghanistan.[99] The first eight pages of *Memorial* consist simply of the capitalized names of all the named combatants
who die in the *Iliad*, printed relentlessly one beneath the other in the order
in which their deaths are described in both the ancient and the modern
poem; the effect is like that of a large inscribed stone war memorial.[100]

In the course of the poem, when each of these men's deaths is narrated,
his name is capitalized again: "IPHITUS who was born in the snow," whose
death is imagined with the aid of one of the most famous similes in the
Iliad in which a dolphin pursues tiny fish.[101] Oswald notes that one of the
virtues of Homer admired by ancient critics was his poetry's visual brilliance, *enargeia*, which she translates as "bright unbearable reality."[102] To
recover this, she strips away the narrative parts of the *Iliad* and concentrates
on just two elements, the similes and the short biographies of soldiers that
often appear at the moment of their deaths. *Memorial*, she hopes, presents
the *Iliad* "as a kind of oral cemetery"; the biographies, she believes, drew
on ancient traditions of lament, to attempt "to remember people's names
and lives without the use of writing."[103]

Oswald is responding to the physical suffering of the Trojan War combatants. Phereclus the Trojan shipwright, for example, died "on his knees
screaming" when Meriones stabbed him through the buttock to the blad-
der.[104] But she also sees that these tiny biographical notices offer us brief

glimpses of the bereaved. The first warrior's death in her poem, as in the *Iliad*, is that of Protesilaus, leaving his new wife in his half-finished house in Greece, "clawing her face."[105]

There are mothers like Laothoë, who will never lay eyes on her son again even as a corpse, since it was washed away.[106] And there are many fathers, such as Antimachus, who put the case for war to his two young sons, "dazed teenagers" who rode to war on their father's pedigree horses.[107]

Dolon's five sisters call upon his ghost at his grave, recalling that he was not much to look at, but fast on his feet.[108] There are men who loved socializing with their friends, among them Axylus from the Hellespont, who died in "a daze of loneliness."[109] Oswald has said that as she reads Homer's obituaries, she can hear under the verse the sound of howling; she intended to pare down the epic to this bleak cry, this howl.[110]

The second major factor in the recent prominence of the *Iliad* has been feminism. Neither Balmer nor Oswald had as a main concern the reappraisal of the *Iliad* from a feminist perspective. But their poems made this point implicitly because women have not historically been much involved with creating receptions of the *Iliad*. They are two of the very few women historically to have felt equipped or daring enough to essay a translation or interpretation of this hypermasculine, solemn, and military masterpiece; the others include, in addition to Weil, an earlier Frenchwoman, Madame Anne Dacier, an exceptional Greek scholar who translated it in 1699, and recently, Emily Wilson.[111] There were already suggestions in antiquity that the *Odyssey* was the product of Homer's weakened genius in old age, and was somehow more "effeminate" than the *Iliad*; Richard Bentley argued that while Homer had composed the songs constituting the *Iliad* to perform at festivals in front of men, those in the *Odyssey* were designed for women.[112] William Golding agreed, saying "anyone who prefers the *Odyssey* to the *Iliad* has a woman's heart."[113]

The masculine focus of the *Iliad* was even more heightened in *Troy*, as it was in John Dolan's savage *The War Nerd Iliad* (2017) and Michael Hughes's *Country* (2018), a clever and prizewinning novel, very different from those discussed in detail here, about militarized male culture; it transposed the action of the *Iliad* to the period of the post-1960s Irish troubles.[114] In the film *Troy*, when Achilles (Brad Pitt) is too busy worrying about Briseis (Rose Byrne) to devote himself to armed combat, Odysseus (Sean Bean) remarks cryptically to him, "women have a way of complicating things." Unfortunately, women scarcely complicate the plot of *Troy* at all. The commanding figure of Hecuba, so prominent in crucial scenes of the *Iliad*, has been entirely deleted. It took Marina Carr's brilliant play *Hecuba*, staged by the Royal Shakespeare Company in 2015, in part inspired by Euripides's *Hecuba*, to reinstate Hecuba to her proper place as the central tragic figure in the entire story of the destruction of Troy after the Greek victory: Carr said she wanted to explore "what a woman's life lived through war was like. What she might have suffered."[115]

In the movie *Troy*, there is no such exploration; its Helen, moreover, possesses not one iota of mysterious power, and Briseis is amalgamated with both Chryseis and Cassandra. Yet alongside the reduction of the *Iliad*'s already meager female quotient, the movie does briefly doff its cap in the direction of its more emancipated third-millennial female audience members by allowing Briseis to stab Agamemnon in the neck. Her action is presented as a feisty postfeminist refusal, when the brutal patriarchal overlord is about to take her captive, to be complicit in her own victimhood. Benioff's otherwise lackluster screenplay here dared—however briefly—to rewrite the Homeric poem in a way that enhances its presentation of female agency. It is an important moment from the perspective of the rewriting of classical texts in modern media.[116]

The first major *Iliad* novel of the remarkable cluster published in the second decade of the twenty-first century was Madeline Miller's *The Song of Achilles* (2012), which won the Orange Prize for Fiction (subsequently renamed the Women's Prize for Fiction). It is a gay love story, told from the perspective of Patroclus, who is in this novel (as in some ancient sources, but *not* including the *Iliad*) Achilles's lover. But it is the non-Greek captive women in the Achaean camp who dominate most of the several Troy novels by women subsequent to *The Song of Achilles*. In 2014, Judith Starkston, an American writer of fantasy fiction, published *Hand of Fire*, a stirring if clumsily written novel taking enormous liberties with ancient mythical traditions and centered on the mystical religious and healing powers of Briseis. Rather improbably, in Starkston's hands, she is madly in love with Achilles. Much more elegantly written is Emily Hauser's *For the Most Beautiful* (2016), which reads as if it were aimed at the young adult market and contains paranormal and fantasy elements as well as romance. It is told through the eyes of both Briseis and "Krisayis," a freely adapted version of Chryseis, who spies for the Trojans in the Greek camp. Hauser, a lecturer in Classics, knows her Mycenaean art and archaeology well, and some of the descriptive writing is excellent.

But it was only when the heavy-duty novelist Pat Barker decided to turn her hand to the Troy story that the *Iliad* found a female fiction writer possessing anything close to the technical mastery, gravitas, and emotional honesty required to do justice to Briseis and the epic. It's rare for a writer who has won prizes for her novels about the trauma inflicted by twentieth-century combat to turn to the late Bronze Age warfare depicted in the *Iliad*. But we should not be surprised that the definitive war epic attracted Barker's attention. She won the Booker Prize for *The Ghost Road*, the final novel in her World War I trilogy, but it was only part of the monumental achievement also constituted by *Regeneration* and *The Eye in the Door*, fiction that equals, in scale and scope, the epic storytelling at which the Homeric poets excelled.

In her earlier novels, from *Union Street* (1982) onward, Barker explored the subjective experiences of working-class women, and it is a commitment to recovering the psychological predicament of powerless women that motors *The Silence of the Girls*, which was shortlisted for two prestigious literary prizes.[117] The setting is the center of Greek military operations, with women enslaved during the sieges of surrounding towns in Asia Minor being brutally herded into what Barker has no hesitation in calling a "rape camp." Her novel's title was suggested by the notorious ancient Greek proverb, "Silence becomes a woman." This was barked by the macho Greek hero Ajax at his concubine Tecmessa in Sophocles's tragedy *Ajax*, where he stabs himself to death after displaying symptoms uncannily close to those which would now be diagnosed as indicating post-traumatic stress disorder.[118] Both Ajax and Tecmessa are briefly brought to vivid life in the book, and the proverb is its leitmotif. Most of the novel features Briseis as first-person narrator, painfully aware of her previous life as the young queen of nearby Lyrnessus. She was taken captive after seeing her city sacked and all her menfolk killed. Her beauty meant that she was spared being thrown to the regular soldiers for their common use. She was awarded instead to Achilles.

As in book 1 of the *Iliad*, Achilles falls out with his supreme commander, Agamemnon, who expropriates Briseis from him. The actions of the subsequent forty or so days, narrated in the *Iliad*, occupy most of the novel—Achilles's wrathful refusal to fight, the desperate attempts of his fellow officers at reconciliation, the deaths of Patroclus and Hector, Priam's visit to Achilles to request Hector's corpse, and Hector's funeral. But Barker also wants us to hear the voices of the women of Troy after it falls, and concludes with material from Euripides's *Hecuba* and *Trojan Women* to expose us to the shock and agony of Hecuba, Andromache, Polyxena, and Cassandra, and to the agonizing burial of baby Astyanax.

The poetry that has informed Barker's whole conception occupies less than twenty lines of the epic, the lines that describe Briseis, once restored to Achilles's tent, weeping over Patroclus's cadaver (19.287–300). For once, we hear her voice, in direct speech, in her address to Patroclus: she tells her dead friend that he had always been kind to her and had promised to make Achilles marry her on his return with her to Phthia.[119] Barker makes us understand exactly the significance of Patroclus's promise. With her celebrated candor, Barker explores the appalling psychological experience of women taken in war—whom Victorian translators used euphemistically to call "spear-brides." After months of terror, when they are finally captured and the rapes begin, the best that any woman can hope for is to be selected by as high a ranking soldier as possible, who will demand exclusive use of her. Briseis is beaten and raped anally by Agamemnon, while sex with Achilles is rough and perfunctory. But anything is better than being forced to service anyone at any time. Exactly the same principle motivates the narrator of the famous 1959 memoir *Eine Frau in Berlin* (A Woman in Berlin), by Marta Hillers: faced with the prospect of being systematically raped by many Red Army soldiers, women in occupied Berlin desperately sought a high-ranking Soviet officer to offer them protection, whatever the favors he demanded in return.

Briseis's account of her ordeals, and those of the other women in the camp, is relentlessly violent. Barker's dialogue is epigrammatic, pungent, and even humorous at times; a whole character can be fleshed out in a single utterance. When the Greek soldiers hand over to Achilles his war prize Briseis—a teenager in catatonic shock who has just watched four brothers die in succession—Achilles just says, "Cheers, lads. She'll do."[120] The uncompromising accumulation of physical details, especially smells, tastes, sounds, and the tactile feel of things, conjures up the squalor of life in the dust, excrement, and gore of prehistoric tent cities. It is only occasionally relieved by a description of the exquisite fabrics the women weave and the tang of olives and fresh honeycakes. The realism is uncompromising: women drink wine, however rancid, as anesthetic whenever they can lay hands on it. They are often as drunk as the men.

The novel's ethical heart lies in Briseis's growing realization that the Greeks are no worse than any other warrior people, and that her only future lies with them. This is not "Stockholm syndrome": it is a pragmatic woman's rational response to a situation in which she is powerless. Because Patroclus is consistently kind to her, she begins to love him, and in due course even accepts that Achilles has certain virtues. At the end, she is not without hope: her will to survive is rewarded. It is its moral sophistication that makes Barker's novel stand out against most of the others based on or connected with the same epic.[121] This applies even to those that, like hers, unflinchingly scrutinize the emotional cost of constant war, such as Malouf's *Ransom* and Miller's *The Song of Achilles*, or put female experience at their center, such as Christa Wolf's much earlier *Cassandra*, however dazzling, and Adèle Geras's lyrical *Troy*, aimed at the young adult market.

Barker's was the most important novel based on the *Iliad* to appear in decades, but just a year later, in 2019, this status was challenged by Natalie Haynes's searing and fiercely feminist *A Thousand Ships*, which was shortlisted for the Women's Prize for fiction the year after *The Silence of the Girls*. What makes Haynes unique as an *Iliad* adaptor is that her book's ambitious scope and grandeur make it seem truer to the vision underlying the *Iliad* than any of the comparable novels. Sublime grandeur was what ancient critics most admired in this epic, and the ancient treatise *On the Sublime*, attributed to Longinus, argues that this was partly due to the extraordinary dignity of the traumatized mortals (9.7): "It seems to me that Homer, in bringing to the gods a range of suffering including internecine conflicts, grudges and vengeance, tears, and bondage during the Trojan War has made his men into gods in terms of strength and his gods into men."[122] Haynes is not nervous about writing words for gods, as are most of the more realist recent novelists; she realizes that the juxtaposition of divine and tragic human perspectives is crucial to the overall elevation of the *Iliad*. She gets round the epic convention of dialogue between gods on Olympus with a sharply scripted chapter on the rivalry between Aphrodite, Hera, and Athena.

Haynes set out to imagine the experiences of numerous females—divinities as well as mortals—directly influenced by the Trojan War. She expands the stories of females who appear only briefly in the *Iliad*, such as Eris, the demigod personifying conflict; or Laodamia, left widowed in Greece when her new husband, Protesilaus, was the first Greek to die at Troy; or Theano, High Priestess of Athena. She also includes females who do not appear in the *Iliad* (Penthesilea the Amazon warrior; Penelope; and Paris's first wife, Oenone), as well as material from several other ancient texts, including the *Odyssey* and Quintus of Smyrna's *Posthomerica*. The first substantial chapter narrates in free indirect discourse the hours leading up to the lonely death of Creusa in the burning ruins of Troy, but from her perspective; our main source for this is Virgil's *Aeneid*, book 2. The tragedies of Euripides also provide Haynes with an important starting point for her several chapters about the relationships between the women of Troy—

Hecuba, her daughters Cassandra and Polyxena, and her daughters-in-law Andromache and (the Spartan) Helen. Euripides's *Iphigenia in Aulis* inspires her painful exploration of the victimhood of this murdered youngster; Aeschylus's *Agamemnon* informs her Clytemnestra at every turn.

But Haynes is clear that her main co-text and the story it tells are epic in every sense of the word. She bookends the novel, which reads like a grand and solemn oratorio for female voices, with the words of Calliope. The cynical muse of epic is tired of Homer's relentlessly male focus and trenchantly concludes, after hundreds of grueling pages, that through her song she has brought back into her reader's consciousness the forgotten women of the Trojan War. Women make up half the people whose lives are touched by war, so we should make an effort to ensure that their experiences are not erased from the record.[123] She has celebrated them "in song." Haynes never forgets that it is a grand and resplendent verbal medium of great tragic force and pathos, originally sung by ancient bards to the plangent lyre, that she is translating into contemporary prose fiction.

Recovering the emotions of these women has been a mighty labor. Nothing in the *Iliad* prepares us for the sense of claustrophobia that Haynes persuasively evokes in describing how women, especially those with small children, must have felt when enclosed within the walls of Troy for ten long years. To compensate for the *Iliad*'s intense focus on friendships between men, she imagines a tender sisterly bond developing between the former queen Briseis and the teenaged tearaway Chryseis; the climax of this chapter is a taut reading of the conflict between Chryseis's father Chryses and Agamemnon, as dramatized in the *Iliad*, book 1, but from Chryseis's terrified perspective.

Haynes points out that the true cost in terms of female fatalities in a war of this kind must have been far greater than the *Iliad* acknowledges, given how many towns in the surrounding Phrygian territories were devastated during the long course of the war. Achilles alone must have killed dozens of women, since some would have refused to accept slavery and concubinage; there would also have been defiant women who died out of loyalty to their husbands or while defending their children.[124]

Haynes's novel, however, finds its sense of tragedy not only in the multiple fatalities, but also in its delicate evocation, reconfigured for our ecological crisis, of that ancient story with which this chapter began: when the Earth asks Zeus to destroy the human race. Her Gaia recalls,

Mankind was just so impossibly heavy. There were so many of them and they showed no sign of halting their endless reproduc-

tion. Stop, she wanted to cry out, please stop. You cannot all fit on
the space between the oceans, you cannot grow enough food on the
land beneath the mountains. You cannot graze enough livestock
on the grasses around your cities, you cannot build enough homes
on the peaks of your hills. You must stop, so that I can rest beneath
your ever-increasing weight.[125]

More recently still, American poet and classicist Alicia Stallings, who has
lived for many years in Greece, has composed a seven-sonnet sequence on
the Homeric epics; the central sonnet draws affecting attention to the brutal, intensive logging in the *Iliad,* especially for the construction of ships:

In rhythms that no bard has ever scanned,
The timber falls. It's timber when it falls
And crashes into silence with its calls
Of birdsong and its rustling sarabande,
A library of turning leaves; its rings
A record of the years no needle traces,
Shade the annihilating sun erases,
Torn from the catalogue of living things.

(It started with the catalogue of ships:
Whole forests felled for keels, masts, spars, oars, hulls
Made black and waterproof with tar and pitch.
The sight of the armada stirred the pulse
Of men more than the hair, the skin, the lips
Of beauty's queen men later called a bitch.)[126]

Perhaps the most important factor in the recent revival of interest in the
Iliad is the sense we increasingly share of a brutal, wasteful, and apocalyptic era that faces the imminent prospect of our entire civilization's extinction. Many of us feel like the herald in Aeschylus's tragedy *Agamemnon,*
who can scarcely believe that he has survived both the Trojan War and the
hurricane that wrecked the Greek fleet on the return voyage: in a reference to the epic tradition with which this chapter opened, he opines that it
is only because Zeus "does not *yet* want completely to wipe out the human
race."[127]

The Greek tragedy most informed by the tradition of threatened total
annihilation, however, is Euripides's *Trojan Women,* used by both Barker

and Haynes. *Trojan Women* is set against the background of the ruins of Troy after the Greeks have successfully besieged it, killed off almost the entire male population, and taken the women and children captive. They are to be separated and deported into slavery. But for Troy itself, of course, there is to be no future: as the chorus sings in the closing dirge, "Like smoke on the wings of the breezes, our land, laid low in war, now vanishes into nothingness." The play's apocalyptic tenor is coupled with metaphysical bleakness. It juxtaposes physically manifest Olympian gods—Poseidon and Athena open the play agreeing to destroy the Greeks by shipwreck as well as Troy in its entirety—and Hecuba's explicit expressions of doubt that the gods can concern themselves with humans or even exist in their traditional form at all. She announces that the gods have "come to nothing" and that all her sacrifices have proved futile.[128]

Euripides found that sense of futility in the *Iliad*'s several unforgettable images of total destruction. The Trojan battalions pour over the bridge led by Apollo. He kicks down the Achaeans' wall "as easily as a child playing on the sea-shore, who has built a house of sand and then kicks it down again and destroys it" (15.363–65). The poet describes how even the last vestiges of the Greek fortifications were obliterated after the war by Poseidon and Apollo, who inundated the shore for nine long days, before Poseidon smashed away every last beam and stone with his trident, made, the coast of the Hellespont smooth again, and covered the beach with sand (12.15–33)—an image that Martin West believes was suggested by the deliberate immersion of the remains of Babylon by waterways that the Assyrian monarch Sennacherib ordered dug from the Euphrates, thus erasing all signs of the city he had conquered and turning its site into a floodplain.[129] The task of imagining proleptically the actual devastation of the regal palace from which he ruled Troy is put in the mouth of the elderly King Priam, when he is pleading with his son Hector not to fight Achilles (22.60–71):

> Father Zeus, son of Cronus, will destroy me on the threshold of old
>   age
> by a terrible fate, after I've witnessed the deaths of my sons and
>   abduction of my daughters,
> my treasure chambers plundered and little children
> cast to the ground in the terrible conflict,
> and my daughters-in-law dragged away at the deadly hands of the
>   Achaeans.

I myself at the last will be torn to pieces at my front gates
by ravening dogs, when someone has taken the life from my limbs
with the thrust or blow of a sharp bronze weapon—
the dogs which I raised in my halls to guard its doors.
They will drink my blood uneasily and lie there at the gate.

For there is a stark tonal difference between the Homeric epics. The Ithacan parts of the *Odyssey*, at least, do not share with the *Iliad* the same celebration of gargantuan consumption; the suitors' profligacy with natural
resources is condemned, and the livestock and orchards belonging to Odysseus and Laertes are on a much more modest scale than those envisaged
in the *Iliad*. Nor is it just that the *Iliad* ends with catastrophe imminent,
whereas the *Odyssey* ends with the victory of Odysseus over the suitors, his
reunion with wife and son, and the outbreak of peace on Ithaca. Nor is it
just that there are several unambiguously comic aspects of the *Odyssey*, nor
that its world is much more open to the possibility of the supernatural and
fantastic than the dogged realism of the *Iliad*.

The Troy epic does, of course, feature epiphanies of gods, one episode
involving talking horses, memories of a centaur, and the account (although
not in the main narrative) of a supernatural monster—the fire-breathing
hybrid lion-serpent-goat, the Chimaera slain by Bellerophon (6.180–82).
Yet the monstrous and supernatural do not predominate. The *Iliad*, which
has no humans who turn into animals, refuses to allow its listener off the
painful hook of war in which humans kill other humans, most of whom are
guilty of nothing. As a deadly serious account of a brutal conflict between
two massive military alliances, it can be transplanted to other situations involving siege, colonialism, or imperialism, but its consistently realist tenor
is not open to the type of surreal, supernatural, and symbolic figures of the
Odyssey's travelogues: the shapeshifting Proteus encountered by Menelaus;
a one-eyed giant; a twelve-footed, six-headed man-devouring sea-monster;
a witch who can transform humans into animals; Sirens; and a journey to
the Underworld. The *Iliad* offers a constrained epistemology.

The history of staging the *Odyssey*, too, reveals that it is difficult to
make tragic drama out of this epic. Most staged versions are humorous.
To turn the *Odyssey* into a tragedy requires altering Homer's presentation
of the hero to make him a much darker, more dangerous character who
ruthlessly destroys enemies and who kills far more suitors than is necessary.[130] When Socrates interviews a professional performer of Homer in
Plato's imaginary dialogue *Ion*, he points to some passages as especially

emotive.[131] All except one are the darkest moments in the *Iliad* (the show-down between Achilles and Hector, and the sorrows of the bereaved An-dromache, Hecuba, and Priam); the only one from the *Odyssey*, revealingly, depicts its hero at his most Iliadic, leaping forth in his true warrior identity to mow his enemies down.

Besides Haynes and Stallings, the other recent women writers re-envisioning the *Iliad* seem at least subliminally aware of its apocalyptic idiom as well. The main part of Oswald's *Memorial* ends with Andromache howling for Hector, returned to Troy "sightless / Strengthless expression-less."[132] But the poem as a whole ends with the word "gone" and an image that connotes apocalyptic cosmic destruction:

> Like when a god throws a star
> And everyone looks up
> To see that whip of sparks
> And then it's gone.[133]

This is a graver revisiting of a famous simile in the *Iliad* when Athena "shot through the sky as some brilliant meteor which the son of scheming Cronus has sent as a sign to mariners or to some great army, and a fiery train of light follows in its wake" (4.75–77). Barker's Andromache, simi-larly, at the end of the novel looks at the deserted sites where the action of the *Iliad* took place; her perception of utter annihilation is more closely associated with the Greek camp than with the Trojan ruins. She imagines winter winds scouring through the huts, saplings taking root, a forest even-tually reclaiming the land, and a beach almost completely empty.[134] The world of ancient Greek literature, of course, reveals as yet no apprehen-sion of a world in which organic life had been wiped out to such an extent that no forest might ever be able to reclaim the battlegrounds of the *Iliad*.

Stallings has heard the loggers at work behind the stories told in both epic poems. The longer works discussed here by Malouf, Oswald, Miller, Hughes, Barker, and Haynes have all been nominated for, or have won, prestigious prizes, but all these writers would probably say that it was Homer who deserved the recognition. Oswald admires the *Iliad* and thinks it worth re-envisioning for a war-plagued planet. But she, like Barker and Haynes, omits the glory to focus on the suffering, resisting Homer's mili-taristic values. Her funereal tone, along with her understanding of the an-cient literary critics' diagnosis of Homer's "bright unbearable reality," re-veals her profound appreciation of the ancient poem's aesthetic achievement.

Her *Memorial,* along with Stallings's sonnet sequence and the very different but equally brilliant novels by Barker and Haynes, shows that the *Iliad*—which ends with women leading lamentations for a civilization on the brink of annihilation—is an ancient poem whose time has truly come.

CHAPTER TWO

The *Iliad* in Its
Historical Contexts

ANCIENT LISTENERS TO THE *Iliad* were invited to create in their minds tumultuous images of apocalyptic environmental catastrophes. They were asked to conjure in their imaginations, for example, a river breaking its banks and obliterating the structures that humans put in place to make the Earth obediently produce sustenance. When the great Achaean warrior Diomedes is running riot on the battlefield (5.87–92),

> He charged across the plane like a river in full winter torrent,
> which instantaneously sweeps away the embankments with its flood.
> The close-fenced banks cannot contain it,
> and nor can the walls of the fruitful vineyards
> when it comes on suddenly, and Zeus's rain falls heavily.
> And many fair works built by human energy are destroyed.

Although representations of wildfire are rare in Greek literature, an exception being Thucydides's references to the fire on Sphacteria in 425 BCE, a berserk warrior can also be likened to wildfire.[1] When Agamemnon slaughters Trojans (11.155–57),

> It was as when annihilating fire falls on a virgin forest
> and a whirling gust of wind blows it in all directions,

and the thickets are utterly destroyed by the force of the oncoming
     fire.

In the preceding woodcutter scene (11.86–90), Jason König observes that
"the poet exposes the precariousness of the anthropogenic image of work
and rest" by portraying "the power of the natural world as so overwhelm-
ing that it drowns out any mention of human presence or human agency."[2]
Yet humans have long contributed to the generation of wildfires around
the Mediterranean, where the characteristic maccia or "maquis" that fringes
it—comprised of combustible nondeciduous bushes, shrubs, and dwarf
trees—is partly a result of human overuse.[3]
     Winds can turn sea waves into lethal phenomena, too. Hector falls
violently upon the Achaeans, and they are terrified (15.624–28):

It was as when a mighty wave, swollen by the winds
high beneath the clouds, crashes onto a ship,
and it's hidden deep in foam; the dreadful blast of wind
roars against the mast, and the sailors shudder in their hearts
through terror.

The geophysical world inhabited by the heroes of the *Iliad* poses constant,
immanent threats to their survival from naturally occurring earthquakes,
floods, fires, and wind-driven sky-high waves, terrifying events with which
the most successfully bellicose of the warriors are also compared. What are
we to do with all this terrifying elemental and meteorological imagery?
     Perhaps it is the product of a generalized anxiety that lay at the foun-
dation of the archaic imagination. The world of the eastern Mediterranean is and has always been susceptible to seismic events, floods, wildfires
and tsunamis. (While I was writing a draft of this book, the tragic earthquake of February 6, 2023, killed tens of thousands of people in Turkey
and Syria; the scenes of utter devastation and human agony are shocking
and unforgettable.) The people living in the eighth century BCE who
produced the *Iliad* in approximately the form we have it knew well that
humans and their handiworks—artificially reinforced riverbanks, vineyards,
ships—were vulnerable to complete destruction by geophysical and environmental forces beyond their control. When archaic Greeks gathered at
festivals of Apollo to listen to itinerant bards perform the *Iliad*, they knew
that earthquake, flood, wind, fire, and wave could threaten their communities at any moment.

Remembering Bronze Age Ancestors

During the Bronze Age and Iron Age, the eastern Mediterranean area where the Greeks lived, especially in and around the Aegean, suffered from frequent volcanic events ultimately linked to the subduction of the African tectonic plate beneath Crete.[4] Particularly significant was the "Minoan" eruption around Santorini about 3,600 years ago, which had been preceded some years before by an earthquake.[5] This eruption may have consisted of a series of three or four seismic events, of which the last was the most destructive.[6] The eruption caused tephra fall, ash fallout, and pyroclastic flow deposits that completely engulfed most of the island, as well as parts of Crete and other islands of the Aegean.[7] These seismic events and resultant landslides generated tsunamis from at least as early as 2000 BCE, as evidenced by inland finds of pumice and shells.[8] Earthquakes were the most important and constant geohazard in the eastern Mediterranean during this time; indeed, central Greece and the Aegean islands lie on multiple fault lines that have continually caused tremors and serious earthquakes over the past four millennia, some of which have resulted in widespread social upheaval.[9] Particular attention has been paid to the wider physical effects and historical impacts of the earthquake that occurred about 1500 BCE and was centered on the island of Kea (ancient Keos).[10]

During the Holocene, human activity became "an increasingly powerful driver of fires."[11] The origins of land degradation certainly extend far back into prehistory.[12] But between around 4000 BCE and the fall of the Roman Empire, what is now generally identified as "Phase 4" in a seven-phase process of land degradation around the Mediterranean began.[13] During this phase, in addition to the livestock grazing of the previous era, there were continued population growth, domestication of fruit trees, extensive clearing of land and terracing of hillsides, and in particular, deforestation for shipbuilding.[14]

As for floods, most of the Mediterranean basin is surrounded by a barrier of mountains that hinders air-mass movements, and when combined with the region's extremely heavy subtropical rainfalls and intense local winds, produces conditions ripe for the development of cyclones.[15] The study of flood history in the Mediterranean and around Crete during the Holocene has been conducted with the aid of landform and sediment-based paleoflood analysis.[16]

But perhaps the post-apocalyptic tone of the *Iliad* is the product of more specific social memory. Pindar reports that King Euxantius of Keos, son of Minos, cited a cataclysm to explain why he preferred to stay in his modest realm with its meager bushes on Keos, rather than moving to a much larger kingdom on Crete, with its cypress trees and "contested" pasturelands.[17] The Cretan cataclysm, caused by Zeus and Poseidon, had destroyed an entire land and army. Richard Janko has made a compelling case that just such a cataclysm was kept alive in the memory of the Kean islanders all the way from Minoan times to the fifth century BCE—for a whole millennium—when traditions about Euxantius were also known to Pindar's Kean rival Bacchylides.[18]

The poets and their audiences were also aware that, several generations before them, the culture that they were reconstructing imaginatively in the *Iliad*, using their inherited genre of epic poetry with all its formulaic language, had disappeared. One response to this was a nostalgic glorification of the past, a sense that society was in a process of decline. Achilles was in a class of his own, able to close the great bolt of his door single-handedly, although it took three of any other of the Achaeans (24.453–56). But men in general were believed to have been much bigger and stronger before and at the time of the Trojan War.[19] This belief may have arisen partly in response to discoveries of petrified remains of the skeletons of immense creatures from the past such as mammoths and mastodons.[20]

Ajax lifts a jagged stone that even the youngest and strongest of mortals of Homer's time could not hold, raises it and smashes it down onto Epicles's head, crushing his skull through the helmet (12.380–85). Hector easily lifts up a stone that Homer tells his audience even two men of their day could not have raised from the ground (12.445–49). Aeneas lifts a stone that would require two of "today's" men to heave up (20.286–87). The *Iliad* even implies that the decline had begun a generation earlier than the action it depicts. Nestor claims the authority of one who had lived in earlier times, among even more warlike men—"Men such as I have never seen since," including Theseus, "a man like the immortals" (1.262–68):

> For those men were the strongest that the earth nourishes.
> They were the strongest and fought most strongly
> with the mountain centaurs, and destroyed them violently.

Nestor had joined them from Pylos, and reports: "No man on earth today could fight them" (1.271–72).

The Homeric poets must have been aware of material remains of these late Bronze Age communities, such as Nestor's palace at Pylos, before the several crises that engulfed them in the eastern Mediterranean from the mid-thirteenth century BCE and throughout the twelfth. Athena smites Ares with an enormous black stone she picks up from the plain, one that "men of former days (*andres proteroi*) had set up to mark the boundary of a field" (21.404–5). Oral memories of these events may have been handed down, both in folktale and formal poetry, though the intergenerational transmission of trauma does not even require explicitly verbalized memories.[21] In the thirteenth and twelfth centuries BCE, the complex, palace-centered communities inhabited by previous Greek speakers in Crete and Greece had ceased to operate at the same time that the Hittite state broke down and its survivors moved south into northern Syria. The cessation is usually framed as a crisis, for example in William Ward, Martha Sharp Joukowsky, and Paul Åström, *The Crisis Years* (1992), and Eric Cline's outstanding *1177 B.C.: The Year Civilization Collapsed* (first edition 2014).

Several different causes of this crisis have been postulated, although the true reason is probably a combination of interrelated factors. Earthquakes and consequent tidal waves have been a popular hypothesis since the 1940s.[22] Human social phenomena were involved, whether as principal causes or responses to natural catastrophe—migrations, attacks by raiders, intercommunity tension, power struggles internal to elites, inequality between social classes or between centers and peripheries. But climatic and disease factors are also likely, which is where the distinction between "natural" and "human" agency begins to break down. Epidemiologists suspect that the large drops in population size must have resulted, at least in part, from a series of outbreaks of a deadly and highly infectious plague.[23] In 1966, Rhys Carpenter argued that there was a drought and consequent famine at the end of the late Bronze Age.[24]

Much more recently, isotope analysis and marine zoology have produced evidence suggesting that, at this precise time, the surface temperature of the eastern Mediterranean dropped by 2°C. Rainfall over land declined sharply, reducing the fresh water released into the atmosphere and creating more arid conditions. The Mycenaeans' palace-centered societies, so dependent on agriculture, would have faced the famine that is named in the *Iliad* (24.532) as one of the worst evils besetting abject mortals and driving them across the face of the Earth.[25] The notion of an unusually wet period followed by an extended drought, one that compelled communities to relocate in search of farmable land, has been supported by studies applying geomorphology, sedimentology, pollen analysis, and radiocarbon dating.[26]

Timber shortfall has also been proposed as a major motor in the Mycenaean decline, especially in Crete. There was once plentiful indigenous wood, available within two miles of the city center at Knossos, as demonstrated by the hoards of shipwrights' tools found there.[27] The story of Minoan and Mycenaean metallurgy near Knossos is one of almost continuous decline.[28] The Minoans "acted as if their supplies of bronze (and fuel) were inexhaustible," but Crete was deforested by the late Bronze Age.[29] Portable braziers, which use less fuel, were adopted for domestic use, and gypsum rather than wood began to be used for doors and thrones.[30]

Much about what caused the changes in eastern Mediterranean society over the century and a half starting in the mid-thirteenth century BCE remains mysterious. But the changes themselves are indisputable. As A. Bernard Knapp and Sturt W. Manning have trenchantly put it,

> The palaces and all their related administrative and economic structures, as well as their representational arts and crafts, came to an end; territorial control dissolved; the Linear B writing system went out of use; international trading contacts dwindled, then disappeared, at least for the time being; and the concept of a supreme ruler (the wanax) and his/her trappings became the subject matter of myth. All this represents a deep rupture in the politico-economic and ideological system that linked together the Bronze Age Aegean world.[31]

A distinct new culture emerged, or rather distinct new cultures. Even within Mycenaean settlements in the Peloponnese there was a huge variety in terms of the nature and scale of the Mycenaean collapse.

The Pylos palace-site was not reoccupied after its destruction. Several of the major centers in the Argolid survived, but signs of human activity in Messenia were wiped out and resumed only after 1000 BCE. Some communities could see and handle the remains of their Mycenaean forebears' civilization, but others could not.[32] In Naxos, distinctively Mycenaean religious practices seem to have survived all the way down to the eighth century BCE.[33] Some Greeks of the eleventh to eighth centuries would have laid eyes on Mycenaean bronze weapons in the form of old votive offerings or in graves that might have been opened: Agamemnon's knife (*machaira*), hanging from his belt beside his sword, corresponds to knives found in Grave IV at Mycenae, which have a ring for hanging them from a belt.[34] Both material objects and social customs as represented in the *Iliad* "can reflect a mélange of different historical periods, including the early Mycenaean, palatial, and postpalatial eras."[35]

Some post-palatial Greeks will surely have seen examples of the Linear B script that they could no longer read, and speculation may have given rise to the story that Proetus sent Bellerophon to Lycia with a folded tablet to give to his new host, one that contained "many baneful and deadly signs" (6.168–69).[36] The kinds of trees present in the Peloponnese also changed. Before the Bronze Age, the region was covered with thick forests of deciduous downy oaks, but these forests were replaced by partially wooded areas dotted with less dense copses or semi-wooded areas (maquis) with evergreen holm oaks and pines. Around 900 BCE, these too gave way, through human intervention, to olive and walnut trees.[37]

Standards of living were generally lower after the decline of the Mycenaeans, but new types of traders and raiders emerged, new tools were devised from a new metal (iron), and new points of contact and the "birth pangs of a new social and economic order" occurred.[38] It was this new order that turned into the society that was ultimately to produce the stuff of the Homeric and Hesiodic epics. The Mycenaean legends, which must be traced back, at least in part, to the rise of Mycenaean civilization in the seventeenth century BCE, were by no means completely forgotten.[39] The new order transmitted the ancient "Mycenaean heroic poetry" that "was cast in hexameters from at least as early as the fourteenth century."[40]

Clues from Mycenaean Greek

Since the decipherment of Linear B and the work of archaeologists and such eminent philologists as Martin West, it has become clear that some elements of Homeric poetry were already in use not only in Mycenaean times but even in the centuries preceding the Linear B records that have been preserved. Some words denoting certain items of arms and armor, for example, refer to items that were already obsolete in the fourteenth century BCE. Some of the shortest similes, consisting of a single phrase, certainly reach back into Mycenaean times.[41] Prepositional preverbs are placed with a freedom that had been lost already in Linear B. And some phrases do not quite fit the meter, but would have done so in earlier stages of Greek.

The society whose palace accountants inscribed records in what we call Linear B has been shown to resemble the world of the *Iliad* in numerous ways. A remarkable fifty-eight recorded names are the same as, or similar to, names of warriors in Homer. Astoundingly, some Mycenaean Greek men bore the names of the top heroes on the Greek and Trojan sides, Achilles and Hector. Other names paralleled in the Homeric texts include Antenor, Glaucus, Tros, Xanthos, Deucalion, Theseus, Tantalus, and Orestes. Sadly, the proper name Nestor has not yet appeared, although many more tablets in Linear B undoubtedly remain to be discovered. A name *ke-re-no* found at both Pylos and Mycenae, moreover, looks similar to Nestor's recurring epithet in the Homeric poems, where he is the "Gerenian" horseman. The only proper name that it may be possible to associate with a historical figure known from other sources is the last king of Pylos, who from Linear B seems to have been called something like "Echelaos."[42] It is enormously suggestive—although it can be no more than that given the present state of our knowledge—that this happens to be the name of the traditional colonizer of the island of Lesbos, far across the Aegean Sea, who was also a son of the Mycenaean mythical hero Orestes.

Linear B confirms the Homeric picture of Greeks to whom sailing and rowing were second nature. Among titles designating occupations, guardians of the coast and shipbuilders receive separate labels. At Knossos, rowers are included in a list of officials supplying or receiving cattle; at Pylos it is possible that some rowers were conscripted, and perhaps the sons of slave women. There is even a specific mention in one Pylos tablet of a naval expedition, since thirty men's names, perhaps the personnel needed to man the oars of a single ship, are designated "oarsmen to go to Pleuron."[43] This is likely to be the city called Pleuron, on the north shore of the Gulf of Corinth, which is named in the *Iliad* (2.639). A likely reason for sailing expeditions, besides trade, was the acquisition of slave labor. Some of the tablets at Pylos indicate that the labor force was supplemented by raids in which women were captured, much like the women taken and exchanged as booty in the *Iliad*. The places where the women are said to have come from are across the sea in the eastern islands and Asia Minor: Lemnos, Knidos, Miletus, and perhaps Chios.[44]

By and large, the gods who have turned up in Linear B are exactly the ones whom we would have predicted from the evidence of Homer and Hesiod. Named recipients of offerings in the Mycenaean tablets are the ones we would expect to be honored by any pagan Greeks—Poseidon, Earth, Zeus, Hera, Athena, Artemis, and Dionysus.[45] Other divinities greatly honored by the Mycenaeans who are named in the *Iliad* include the childbirth-goddess Eileithyia and the Winds.[46] The offerings the Mycenaeans' gods receive are recognizable from Homer: they include cattle, sheep, wheat, barley, oil, and wine; offerings of a non-edible form include sheepskins, wool, and a golden cup, as well as at least one woman.[47]

The women in the palaces worked at carding wool, spinning, and weaving, while both men and women seem to have been involved in making clothes and working flax, which would also have been crucial for equipping ships with sails and both fishermen and hunters with nets. Women ground and measured grain, but men made the bread. Male stokers and ox-drivers, and female bath-attendants and serving maids, are also attested.[48] Occupations reminiscent of the personnel of the *Iliad*, in its descriptions of action and artifacts as well as its similes, include shepherds, goatherds, fullers, priests and priestesses, messengers and heralds, bronze smiths, goldsmiths, bowmakers, cutlers, huntsmen, woodcutters, masons, shipbuilders, and carpenters.[49] Named timbers include elm, willow, and cypress; furniture is decorated with kyanos, horn, and ivory.[50] Horses are mentioned, but not often, which implies that they were used for chariots rather than plows and farm carts; deer and asses make appearances, and dogs are implied from the first element in the word for huntsmen, *kun*-agetai.[51]

The *Iliad* looks backward, so it contains traces of poetic language and memories of the Bronze Age, alongside some conjectural awareness of the complex causes that brought the palatial societies to an end. The poem may have been repeatedly updated, and expanded to account for new experiences, conflicts, and natural disasters that happened in the Mycenaeans' last two centuries, simultaneously with or shortly after their occurrence, and elaborated by the "Dark Age" communities that followed them. Who knows when the images of earthquake, flood, wind, and fire first appeared in the Greeks' ancestral poetry? We do know that when the Mycenaean civilization arose, which two or three centuries later was to take over Crete, the Minoans on that island were recovering from a terrible earthquake that marks the transition from what archaeologists call their First and Second Palatial periods.[52] Certainly the images of total destruction by quake, fire, flood, and wave that we see today were not unfamiliar to those earlier generations. And we are witnessing such scenes on our television screens almost nightly, as thermometer readings hit ever higher numbers.

Transmission through Greece's "Dark Age"

In Euboea, the long island stretching parallel and close to the southeast coast of Greece, the excavations by the British School at Athens at Xeropolis, which may be the original site of the city of Eretria, have shown that it was inhabited continuously from the Mycenaean period to the eighth century and beyond, including the so-called dark period from 1100 to about 750.[53] Xeropolis therefore raises vital questions about the transmission of culture—especially the heroic poems and the gods and myths they celebrated—from the time of Nestor's palace and the Mycenaean takeover of Crete to the introduction of the Phoenician alphabet. There is evidence that Mycenaean objects were handed down as heirlooms over the centuries. At both Eretria and Lefkandi there are several cases of possible heirloom funerary offerings, such as a Mycenaean bronze spear head, a woman buried with a Babylonian gold pendant (dated as early as 1700 BCE), and Mycenaean beads and bronze jugs.[54]

The most arresting item that these Euboeans have bequeathed to us is a clay statuette of a centaur, over a foot high, exquisitely decorated with a geometric dog-tooth pattern. The centaur's head and body were found in separate tombs, suggesting that they were so precious to two members of the same family that mourners broke the piece to bury parts with each. These people were fond of one another. The centaur is hollow, was made using a potter's wheel, and dates from the tenth century BCE.[55] Although tenth-century images of centaurs have been found in Cyprus, none of them shares the same quality of manufacture and design. With this centaur, we are looking at something much loved by a tenth-century Euboean family that did not feel they were living in a "Dark Age." I suspect that they already knew from their myths and poems that the first medical doctor was a man-horse, Cheiron, "the wisest of the Centaurs, teacher of Achilles," as he is in the *Iliad* (11.832–34). We shall see later that, besides Achilles and the centaurs, some of the aspects of the *Iliad* discussed in this book, for example its obsession with highly wrought metal artifacts, may have been a feature of narrative poetry already in Mycenaean times.

The *Iliad* emerged in something like the form we have it in the eighth century BCE. But its Mycenaean elements must have been supplemented over the intervening three centuries, when Greeks, as they adapted to their new situations, continued to tell each other ancestral stories. In the first phase of the hexameter poetic language, many features of the Aeolic dialect of Greek were absorbed, before this poetic diction's final, Ionian phase.[56] Around 1050 BCE, the inhabitants of the villages that would eventually coalesce into the port city of Corinth could all meet each other at Isthmia, where the Peloponnesian peninsula meets the mainland, to sacrifice in Poseidon's honor. While there, they must have told various stories about their ancestors, their voyages, and their wars, combining memories handed down over the generations with creative fantasy. It is unsurprising, then, that they developed two contrasting mythologies about the deep human past.

One was a positive account of the "rise" of human society through the gradual acquisition of technology, cultural practices, and increasingly refined products, a narrative later seen in the Protagorean and Aeschylean account of Prometheus, and in Diodorus's narrative of human history.[57] Aristotle describes distinct "lifestyles," or what we would call "modes of production of livelihood," *bioi*, as hunting and fishing, raiding, and pastoralism, but he was describing different human societies contemporary with his, rather than evolutionary stages. His Peripatetic follower Dicaearchus of Messana, however, articulated in his lost third-century *Life of Greece* three successive stages for how humans acquired sustenance: gathering, then pastoralism, then agriculture.[58]

Set against these narratives of progress was the second narrative, one of systematic decline. Consider, for example, the series of apocalyptic annihilations of all humankind articulated in Hesiod's myth of the successively inferior ages of man in his *Works and Days*. This text has been described by Mark Payne as an early example of "apocalyptic fiction," although Astrid Möller argues that Hesiod's picture of humans' future is not entirely bleak, provided that we follow his stern moral and agricultural injunctions.[59]

Hesiod tells us that the blissful golden race of men was succeeded by the naive and irreligious silver race. Third came the bronze race, terrible and powerful, obsessed with fighting and wanton violence—"the grievous works of Ares"; they did not eat bread and had no iron.[60] Their armor, houses, and tools were all made of bronze. They were destroyed at their own hands and left no name behind them. The fourth generation to appear was a "godlike race of hero-men, who were called demigods (*hēmitheoi*), the race before ours across the boundless (*apeirona*) earth."[61] War and battle destroyed some of the men across the "boundless" earth at Thebes and at Troy, but some were translated to a serene afterlife at the ends of the earth by the shores of Ocean (the circular sea that surrounded the entire Earth) in the islands of the blessed.[62] Hesiod's own generation, the fifth, is the race of iron, doomed to incessant labor, misery, and death. Strife and dissension in families and communities run rife; there are perjury, envy, siege warfare, and impiety, while honor is paid to unjust, violent men.[63]

Although Hesiod explicitly identifies the generation of men who fought and died at Troy with the race of demigods, his audience knew that the heroes of the *Iliad* were actually an amalgam of all three of the last five races. These warriors do eat bread and do not normally live in bronze houses, but they are outsized and addicted to warfare. They use profligate amounts of bronze as well as iron, but wield, with two exceptions (Pandarus's arrowhead and Areithoüs's mace, discussed later), only bronze weapons; they conduct siege warfare, admire violence, suffer from acute inter- and intracommunity strife, feel envy, and commit perjury and impious acts.[64]

The archaeologists' discovery of the Euboeans, from the tenth and ninth centuries BCE, reminds us of the association between Euboea and Hesiod, who tells us that he once sailed to Euboea and went to Chalchis (the other major town of Euboea besides Eretria), where games were being held in honor of the deceased leader Amphidamas. Hesiod tells us that he won the competition between singers, and the prize of a tripod with handles.[65] How far back in history had the Euboeans been holding such contests? Bards could have been competing in Euboea, performing early versions of parts of the *Iliad*, through all the long centuries since the Mycenaeans.

It was also during the supposedly dark eleventh, tenth, and ninth centuries BCE that several important cities were founded on the coast of Asia Minor, in what is now western Turkey. A wave of settlers arrived by sea from regions of mainland Greece, including Euboea, Phocis, Thebes, Athens, and the Peloponnese. This movement eastward is conventionally labeled the period of Greek "migrations" rather than "colonization," to differentiate it from the much larger-scale expansion across the Mediterranean and Black Sea that ensued in the later eighth century BCE.

If we knew more about life in the Ionian cities of Asia in the tenth and ninth centuries BCE, we would be in a better position to understand how the world imagined in the *Iliad* relates to history. Cultural interaction with the ancient peoples encountered by the Greeks in the east must have played a crucial role. Greek poetry may at this time have begun to absorb the extensive motifs it shared with Ancient Near Eastern mythology and literature. Relationships with peoples who appear as Trojan allies in the *Iliad* were sometimes cooperative. Later interactions with the Carians involved intermarriage and Herodotus said that the residents of Miletus spoke Greek with a Carian accent. It may have been from the Lycians that the Greeks learned to worship Apollo, as the god's epithet *Lycian* suggests.[66]

Homer's *Iliad* created for the Aegean Greeks, west or east, a picture of their warrior forefathers. It provided them with a detailed narrative about the Greek-speaking men of the heroic age and how they had sailed over the Aegean to fight the indigenous peoples of Asia and wrest property and women from them. The catalogue of Achaean ships in the poem is a roll call, designed for an eighth-century audience, of the twenty-eight contingents of Greeks, in over a thousand vessels, who participated in the Trojan War centuries earlier. The catalogue must contain much older, inherited Mycenaean material, but it was given something like its present form *after* the Greek migrations to Asia, which must interfere with the way it constructs the distant past.

The "mindset" expressed in the *Iliad* paradoxically combines human technocratic arrogance, terror of natural forces, and metaphysical confusion about the role the gods play in activating these natural forces. The deep history of the poem means that we cannot precisely pin down the mentality of the poem to any one century, let alone a particular place and time. But the whole temporal period over which the *Iliad* evolved coincided with increasingly swift deforestation, and has been posited by some as heralding the onset of the Anthropocene, which makes it a particularly important text given our current urgent imperative of considering whether the Anthropocene has any kind of future.[67]

CHAPTER THREE

Defining an Ecocritical Approach

THE *ILIAD*, ONE OF the earliest Greek poems, evokes a society engaged in profligate consumption and the destructive cause of siege warfare. Might it serve as a warning, helping us to stop our abuse of the Earth and instead apply our brains, our inventiveness, our human culture, to reintegrate ourselves into the natural world? As Jason König has recently commented, it seems odd that the *Iliad* has "not had a higher profile in the history of ecocritical thinking, or been used more widely as a resource for creative responses to the theme of human-environment interaction in the present."[1] His remarkable book focuses on mountains in the ancient world. In the *Iliad*, images of mountains offer a juxtaposition between a "claustrophobic, immersive version of human-environment interactions" and an "extravagant, exhilarating celebration of human control." This makes these images a powerful "imaginative resource against which to measure our own similarly conflicted experience" of our relationship with the natural world we inhabit.[2]

James Redfield's pathbreaking anthropological study *Nature and Culture in the "Iliad"* has informed my thinking in this book at every point. I have also been deeply influenced by environmental historians such as John Perlin who have suggested that ancient societies caused their own collapse by destroying supportive environments, for example by deforestation. As Perlin explains in his *A Forest Journey: The Role of Wood in the Development of Civilization*, the *Iliad* can be read as exposing the dangers inherent in the routine deforestation initially undertaken to acquire timber for housing, heat, shipping, fortifications, funerals, and watchfires, as well as the further deforestation required to clear pastoral land for the grazing of livestock, especially cattle. These were consumed in the Homeric poems with great abandon in feasts and sacrifices and exchange rituals. The timber needed to fuel the forges that produced the metal objects, especially arms, displayed in such extraordinary excess in the *Iliad*, was also a huge factor in the rapid deforestation of the littoral of the eastern Mediterranean, an abuse from which it has never recovered.

In *Landscape and Memory*, Simon Schama writes of "cultural reaffor-estation," where the "literary and visual imagination" works to "replant" diminished woodlands—Shakespeare's plays, enacted within wooden the-aters, offer the Forest of Arden at a time when British deforestation, driven by the need for ever more ships to support incipient colonialism, was be-coming irrevocable.[3] In the *Iliad*, there are a very few attempts to imagine a harmonious utopia in which humans exist within nature, for example in the convention of nostalgically pastoral scenes where heroes are recalled herding flocks on hillsides; it is implied that they did so before aggressive woodland clearance for pasturage had taken place. There is some inter-connection between local and global scales, as König points out; the in-terconnection is conveyed especially through the networks of similes re-quiring the audience to visualize numerous other, socially and geophysically different, locations. Their "abrupt switches of location and perspective are precisely the kind of shifts that modern ecocritical discourse tends to value so highly for their ability to make us aware of the global scale lying behind local environmental phenomena."[4] Yet, in the *Iliad*'s main narrative at least, there is little more attempt to imagine a more harmonious relationship between people and the environment than there is to describe more peace-ful interactions between people of different social classes and military con-federations. The loggers' axes resound cacophonously at every turn.

As discussed in Chapter 2, the Homeric epics have been rediscovered over the past decades by not only the academy, but also the public at large, for their ability to speak to modern ecological concerns. In 1988, John W. Head, professor of law at the University of Kansas, published the third and final book in his commendable trilogy on the need for radical reform of farming in the face of ecological degradation—*Deep Agroecology and the Homeric Epics: Global Cultural Reforms for a Natural-Systems Agriculture.*[5] Head opens with two questions: "Our eyes widen when we see such names as *Iliad*, Odysseus, Achilles, Moira. What, after all, does the return of Odys-seus from the Wars of Troy have to do with modern agriculture? . . . Maybe we need a fresh look: Might that fresh look come from 2,800-year-old ideas?"[6]

Head frames his prescriptions for legal guarantees for sustainable agriculture as his own *aristeia* in the face of the enemy threatening the world with environmental crisis. He imagines himself as a Homeric warrior channeling his wrath against a corrupt system of values, "giving special attention in our case to those values that have built an exploitative system of extractive agriculture."[7] If we fight like Achilles against the vested interests that make agroecological reform impossible, we could win imperishable glory down the generations to come for our part in reversing environmental degradation.[8] By harnessing our intellects and cunning, like versatile Odysseus, we can change our fate and achieve a homecoming that returns us to sustainability.[9] As humans we are mortal, but as a species, he assumes, we are like the immortals, and have a duty of trusteeship toward the planet.[10]

Head's convictions now seem prescient, although these days his adoption of the Homeric hero's bellicose values seems inappropriate to the environmental cause, and his assumption that as *Homines sapientes* we are collectively deathless seems laughably optimistic. Classical literature itself is now only around three millennia old. But Henry Gee, senior biology editor of *Nature*, calculates that if we carry on the way we are, we have only about five hundred years before the planet will undergo the next Great Extinction and humans will be no more; even if we mend our ways soon, the longest we have before we vanish entirely is likely a very few thousand years.[11]

In twenty-first-century poetry and fiction, we can begin to sense that the epic's sense of impending doom is connected with the abuse of nature as well as cavalier attitudes toward battlefield fatalities and the rape and enslavement of women. It is surely time for a new close reading of the *Iliad* that will put our changing environmental consciousness at the center of our critical radar, and so, as Gillian Rudd writes, "bring to the surface elements which are habitually overlooked."[12]

Classical material has always been, and always needs to be, open to new readings. Richard Martin has brought the speeches of the Homeric epics to vivid new life, applying recent techniques from ethnography and sociolinguistics to show that Homer represents an actual form of speaking in a traditional culture.[13] Since the 1970s, if not earlier, the relentlessly patriarchal narratives of ancient myth and history have been systematically retold from the female viewpoint, by novelists and playwrights as well as scholars.[14] Anticolonial critiques and interpretations of ancient attitudes to ethnicity and slavery have fueled intense creativity in anti-racist literature, art, and academic studies since the 1930s.[15] Queer writing about and in response to antiquity is now abundant.[16] The lost voices of classical Athenian teenagers and young adults, destined for the battlefield or dangerously early childbirth, have been reclaimed.[17] Even social class—long an "ugly sister" within the progressive field of classical reception studies—is beginning to be treated seriously by scholars, opening up possibilities for more class-conscious reworkings of ancient texts.[18] The myth about three disabled divinities—Cedalion, Hephaestus, and Orion—cooperating on Lemnos has been repurposed by scholars striving to use disability studies to deepen our understanding of ancient representations of the body.[19] If works of canonical literature can be reinterpreted from feminist, anti-racist, postcolonial, pro-youth, and anti-classist perspectives, and by those critical of prejudice against disability, why not, in the twenty-first century, reimagine them from a subject position that acknowledges the new extent of our environmental terror?

Scholars and creative artists have proposed different metaphors for the process of such repurposing. Theater director Peter Sellars sees each classic text as an antique house that can be redecorated in the style of any era, while remaining essentially the same.[20] Oliver Taplin proposes the more volatile image of Greek fire, a substance used as a weapon that burns under water. Greek culture, according to this analogy, is still present in invisible yet incendiary forms.[21] More ambivalent is Derek Walcott's description, repeated in poems including *Omeros*, of "All that Greek manure under the green bananas"—the Greek legacy is excrement, but it has also fertilized his Caribbean imagination.[22] This phrasing captures beautifully the paradoxical nature of ancient Mediterranean discourses to people who have lost so much by being colonized by western powers. Tracking the ethos of human arrogance toward nature in the *Iliad* may at times be uncomfortable, but the process may help in a small way to foster change.

Alongside the rampant destruction of forests that the way of life—and death—portrayed in the *Iliad* assumes, there are also a very few passages that reveal a greater human humility toward nature, disclose at least a dawning awareness of humans' abuse of it, or celebrate its (vulnerable) beauty. Some environmentalists insist that the ancient Greeks did hold superior views to ours on the importance of respecting wildlife and unspoiled nature, pointing to the figure of Artemis, for example, as protectress of fauna and flora: "Artemisian wildlife refuges could not have been better located if a modern land manager had chosen them to represent each Mediterranean ecosystem."[23] M. D. Usher, a professional classicist and self-taught agriculturalist, has argued that, rather than lambasting the

ancients for their insensitivity toward the environment, we should investigate the not-negligible conceptual roots in ancient thought and culture of
sustainable systems.[24] Alfred Siewers has done the same in an attempt to
rehabilitate Genesis as a foundation text of human rapacity toward nature,
preferring to recover signs of ecological sensitivity in this other premodern text.[25]

But acknowledging signs of both human destructiveness and ecological awareness simultaneously is to acknowledge Pierre Vidal-Naquet's observation that ancient literature transcends history because it crystalizes
contingent ideological conflicts, challenges our values, and is therefore unusually susceptible to diverse interpretations.[26] Raymond Williams would
have suggested that this was in turn made possible by the ideological complexity of the original epic, according to his notion that any moment in
time contains three strands of ideology: old-fashioned ideas on their way
out, dominant ideas that the majority of people hold, and emergent ideas,
developed only by avant-garde segments of the population, that may not
become mainstream for centuries.[27] On this argument there are things in
the archaic *Iliad* representing emergent ideologies that might not become
prevalent for millennia. Examples might include the demonstration of the
unfairness of inherited power over meritocracy (in the conflicts between
Agamemnon and Achilles in book 1 and Odysseus and Thersites in book 2),
or the demonstration in book 21 that if man damages rivers there will be
consequences that are damaging to man.

Williams's helpful concept of emergent ideology approximately corresponds with the Russian formalist Mikhail Bakhtin's more obscure notion that literature holds what he calls "prefigurative" meanings that can
only be released by reassessments that will occur in a distant, "great time"
in the future.[28] An ancient text can "prefigure" a far later concern if we
look at the history of literature as a whole from our later vantage point. Yet
another way of putting this is Erich Auerbach's mystical concept of "figura"
or "umbra," which draws on medieval allegorists to develop a metaphor of
literary "prefiguration" or "foreshadowing."[29] According to this argument,
an element in an ancient text (for example, the *Iliad*'s rampant logging expeditions on Mount Ida, which were necessitated by a war imposed by an
aggressive and acquisitive overseas military power) can in a mysterious but
profound manner prefigure or foreshadow things that happen later (the
twenty-first-century threat to the last remaining woodland of Mount Ida
posed by overseas gold-mining interests).

Ultimately, it does not matter which theoretical concepts we use to describe the repurposing of ancient literature to explore contemporary concerns. I agree with the robust manifesto of Jeffrey Cohen and Lowell Duckert in their edited collection, inspired by Empedoclean thinking, on the classical elements of air, fire, water, and earth in cultural history. We are in a position where we "cannot see the trees for the deforestation."[30] We need to be anachronistic, if necessary, if that anachronism is constructive and productive. When it comes to the state of the planet, we have a collective amnesia, labeled by Ghosh our "great derangement," by which we license devastation.[31] We need to reactivate "lessons from the past . . . for better futures"; history offers "a storehouse of imaginings in which nature is understood as an active force, unlooked-for partner," and an "archive of irremediable precarities."[32] The *Iliad* offers all these things, and more. But environmentally sensitive ways of interpreting literature are still in their infancy. What analytical tools can we bring to bear on our project of reading Achilles in green?

Materialism and the Environmental Humanities

In the past few decades, both outside and within classical scholarship, new initiatives in ecological thinking about human literary culture have emerged, some of which have stimulated the questions asked in this book and the methods used to try to answer them. The least useful, unless handled carefully, has been the fashionable so-called new materialism, with its popular terminology of "unruly matter," "vital matter," and "thing-power." New materialism is a label applied to a range of intellectual approaches across several disciplines, including philosophy, anthropology, political science, environmental studies, and cultural/literary studies.

But one aspect of new materialist literary analysis that we would do well to resist is its retreat from the relationship between *work* and matter—what Marxists call the "labor theory of value." As James O'Connor trenchantly put it in 1998, since environmental history is "the history of the planet and its people and other species' life and inorganic matter insofar as these have been modified by, and have enabled and constrained, the material and mental productions of human beings," the relationships between humans and non-humans are completely "indecipherable without an investigation of the social relations between human beings ('society,' 'economy'), as well as nature's own biological, chemical and physical relations."[33]

The word "materialism" presents a specific problem for scholars because, like other philosophical terms such as "hedonism," it currently means different things inside and outside the academy. Hedonism, for the general public, means an excessive love for physical pleasures, but for scholars of ancient philosophy it means cultivating peace of mind according to the precepts of the Epicurean school. The two things that materialism designates are not only very different, but often perceived as virtually antithetical. Materialism, to at least 90 percent of people speaking English and many other languages, means an attitude toward life that prioritizes the pleasurable consumption of material goods, along with the accumulation of possessions and the wealth that can provide them. It is a word often used pejoratively by people who prioritize other life goals such as intellectual development, spirituality, or social justice. In popular political language, it is broadly denigrated by "left-wing" individuals as a characteristic of "right-wing," conservative, pro-capitalist people.

By contrast, within the academy, since the early eighteenth century, materialism has meant a scientific and/or philosophical system, traceable back to Democritus, that is often atheistic and socially progressive and that asserts the primacy of matter and regards what we call "god," "mind," "spirit," "consciousness," "transcendence," and "ideas" as fundamentally products of—and as caused, conditioned, or informed by—material or physical agencies. By the late 1880s, the word "materialism," even unqualified by an adjective, began to be identified, more and more exclusively, with the revolutionary political philosophy of Karl Marx (whose doctoral thesis was on the subject of ancient atomism) and Friedrich Engels.[34] This shift was a result of the coining of the labels "historical materialism" and "dialectical materialism" to describe their historical and philosophical methods of enquiry.[35] These methods situated humans as organic beings in constant interaction with other organisms and their material environments; they envisaged the nature of human consciousness as culturally and historically relative precisely because it is informed by these interactions, especially those related to the production of goods necessary to survival.

In most of human history such production has entailed an enormous amount of human labor, marked by conflict between poor laborers and those non-laborers who grew rich on their productivity. These different groups can be defined according to their relationship to the production process, that is, according to their objective "class." The word "materialist" in this technical, philosophical sense, partially overlapping with the term "Marxist," is often found doing the opposite ideological work from that which the "consumerist materialism" does. It is used by "right-wingers," often religious conservatives, to attack egalitarians and socialists: it is said to be a "dogma" that reduces the status of the human individual to that of a "mechanical automaton."[36]

This bifurcated signification of the word "materialism" and its use in adversarial polemic should alert us to the possibility that the emergence of new materialism is doing ideological work of a political nature, however hidden any agenda may be. Many of the most prominent new materialists, for example Maurizia Boscagli in *Stuff Theory: Everyday Objects, Radical Materialism*, actually go out of their way to position their arguments as a refutation, or at least an adversarial rival, of Marxist cultural theory.[37] One of the most cited of the new materialists among classicists, Jane Bennett, has openly and explicitly differentiated her own understanding of the M-word from that of intellectuals working in the historical and dialectical materialist traditions.

> I want to emphasize, even over-emphasize, the contributions of non-human forces . . . in an attempt to counter the narcissistic reflex of human language and thought. What counts as the material of vital materialism? Is it only human labour and the socioeconomic entities made by men and women using raw materials?[38]

The intellectual wriggling here is complicated. Human subjects need to be downgraded in our appreciation of matter. Matter and objects have a vitality, instrumentality, and, it is implied, an almost conscious agency of their own. We humans are narcissists—cosmic imperialists who by imposing "subject"/"object" hierarchies somehow oppress inorganic elements, minerals, liquids, and gases as well as organic flora and fauna, at least if we do not acknowledge their immanence and vitality. And "human labour" and "socio-economic entities" have, Bennett implies, unfairly monopolized the attention we humans pay to matter. She continues:

> Or is materiality more potent than that? How can political theory do a better job of recognizing the active participation of non-human forces in every event and every stabilization? Is there a form of theory that can acknowledge a certain "thing-power," that is, the irreducibility of objects to the human meanings or agendas they also embody?[39]

I must stress that I have absolutely no objection to questioning anthropocentrism. The ecological crisis that we *Homines sapientes* have inflicted on planet Earth demands that we change our exploitative and destructive behavior toward all the "things," animate and inanimate, with which we share our Earth and the universe.[40] But there is a fundamental flaw in Bennett's premise that this exploitative and destructive history is connected with thinking about matter *exclusively* (her word is "only") in terms of labor and socioeconomics.[41]

It needs to be countered that we have never yet paid remotely *enough* attention to the relationship between material things, human labor, and socioeconomics. We can surely *add* some of the language of "vital materialism" to our interpretive toolkit when working within any academic discipline. But the idea that any academic discipline has already done a good enough job of thinking about labor is preposterous. Only a scholar working in a country like the United States, where only about 20 percent of the workforce are engaged in agriculture or industry while the other 80 percent operate at an extreme degree of alienation from the processes of material production, could possibly hold such an opinion. In some European countries the productive personnel now constitute less than 5 percent of the working population. But try claiming that we are too focused on labor and the socio-economy to a citizen of Zambia or Burundi, where the percentage of the workforce laboring in agriculture or industry is 91 percent and 96 percent, respectively. Globally, more than 40 percent of the workforce still works in farming of one kind or another, often at a subsistence level, with scarcely any machinery, and in abject poverty.[42] Every year, the number of humans involved in industrial labor worldwide increases.[43]

Bennett's aversion to labor-centered discussions of matter reminds me of Joseph Wright, the illiterate workhouse boy and wool sorter who eventually rose to become professor of comparative philology at Oxford, but remained proud of his origins all his life. His wife, a young woman from a much more privileged background whom he had met when she was studying at Lady Margaret Hall, recalled a rare occasion on which he had rebuked her. She had facetiously complained that doing philology, and consulting big dictionaries, required excessive "manual labor." Wright quietly pointed out that "manual labor" meant working with, for example, a wheelbarrow.[44]

One reason for introducing these global and cross-class perspectives on new materialists' suspicion of the role of labor and socioeconomics in thinking about matter is partly my irritation at the feeling expressed by some of them that they are occupying higher moral and more radical political ground than the rest of us anthropocentrists. But the other is that the society which produced the Homeric epics was, in terms of its relations of production, far more similar to modern Zambia and Burundi than it was to modern England or the United States. If we are to appreciate fully the role of materials and objects in poems composed between the late Bronze Age and the late eighth century BCE, then we surely would be well advised to ask how those materials were thought about in that society, as well as their "vitality" or "thing-power."

Rather earlier than new materialism, a concept emerged among psychologists in response to the environmental crisis. "Ecological consciousness," as they call it. This links the parlous state of the planet with the "bounded, individualistic, model of the self that has dominated Western cultures for at least the last two hundred years—itself closely related to the rise of scientific rationalism and materialism."[45] Seeing individual consciousnesses as individuated and separate "islands of consciousness, disconnected from one another and from the rest of the world around us," has certainly "contributed to our growing sense of disconnection from the natural world."[46] This in turn has played a role in ecological degradation through our mindless exploitation of natural resources.[47] Jung's model of the mind has been found especially suitable because he assumes that myths not only belong to the ancient past of their creation, but also actively map what is fundamental in our culture today: they lay the foundations of our consciousness.[48]

Such psychologists argue that we need to encourage the emergence of a more ecologically sensitive selfhood in which we include the environment in our self-definition. This would encourage us to realize that conservation of the natural world is at the same time a practice of self-preservation, as concluded in Freya Mathews's landmark study *The Ecological Self.* The notion of "ecopsychology" has emerged, deriving from Theodore Roszak, who argued that ecology is a matter for the unconscious mind as well as the conscious sense of selfhood: the very core of our minds is constituted by our "ecological unconscious." He saw this as the living record of what he calls "cosmic evolution," which can be traced back to very early human thought patterns—an unconscious that connects us to all other life on Earth.[49]

Labor and Ecology

The term "ecological unconscious" is also suggestive of the idea of the "political unconscious," a label coined by the great American Marxist lit-

erary critic Fredric Jameson, who laid out his concept in his 1981 book
The Political Unconscious: Narrative as a Socially Symbolic Act. As a Marxist,
Jameson believed that history has been driven by the conflicted relationships between human classes, which can be defined by their basic relationships with the means of production—land, factories, farms, mines, forests,
seas—and the work they do or do not do within the productive process.
Since all art and cultural artifacts, including literary texts, must be tied
in some way to the societies that produced them, they can be analyzed in
ways that are sensitive to and expose the form taken by those ties. The
function of the culturally produced text is to reconcile in the aesthetic
realm the conflicts that cannot be reconciled in the world of historical,
materially lived, reality. If we analyze a text thoroughly we can, argues
Jameson, discover "symptoms" and clues to the political unconscious underlying it. The process by which class conflict is transmuted into art, as
identified by Jameson, resembles in part the aestheticizing function of art
identified several decades earlier by Walter Benjamin in "The Work of Art
in the Age of Mechanical Reproduction."[50]

The literary author and his audience are usually not fully conscious of
these connections, and may avoid altogether explicit discussion of topics
such as labor, power structures, and wealth. But even evasion, whether an
unconscious denial or conscious avoidance, is a political act. A famous example cited by Henry Louis Gates Jr. is the genre of the Confederate romance, in which the nineteenth-century Southern plantation is presented
as a place of conjured delicate sentiment, magnolia blossoms, and moon-
beams—images designed to obscure or eradicate the reality of systemic
slave exploitation, rape, and torture that made the entire economic system
of magnolia-scented plantations possible in the first place.[51]

The *Iliad*'s near-total erasure of the labor that would be required to
service the needs of the warriors is an obvious equivalent: the hard domestic labor carried out by females that is clearly attested in the Linear B records, and the female sex work that it is vaguely implied took place in the
Iliad's Achaean camp, are either edited out or reduced to glamorous cam-
eos: we hear of a millstone in a simile (7.270), but of no armies of women
incessantly grinding grain. The aristocrats Helen and Andromache, in
their luxurious palace chambers, weave scenes from the Trojan War and
pictures of flowers at their respective looms (3.125–28, 22.440–41), but
these bravura images of solo artistic production are somewhat undercut by
the mass labor implied by the huge number of loom weights and spindle
whorls found in the late Bronze Age ruins of Hissarlik.[52] Other soft-focus

images include the lovely-haired, goddess-like Hecamede preparing a beverage in a gorgeous and expensive vessel (11.623–41); a distant memory of beautiful Trojan women doing their shining laundry in attractive stone washing pools in their local geyser (22.153–56); and a tiny handful of similes depicting women working in the luxury ivory or textile industries (4.141–44; 12.433–35).[53]

Agricultural labor features only briefly in similes and on Achilles's shield; mining, ore processing, and transportation are erased altogether. The basic word meaning "exertion" or "labour," *ponos*, usually refers to what warriors do on battlefields, such as fighting over Patroclus's corpse (16.568). Indeed, *ponos* can stand alone as a metonym for "battle" (6.77, 6.525, compare 13.2, 17.158). Much less frequently, it is applied to other kinds of physical exertion, such as cutting up sacrificial meat into joints (1.467) or undergoing emotional distress (21.525).

Among *Iliad* scholars, the deep transdisciplinary model of the type of criticism that intelligently probes the relationship between social/material/ political/economic reality and its aesthetic projections can be felt, for example, in Peter Karavites's excellent observation that when poetry is great, "it is so precisely because it is creative, which means (among many other things) that it has exaggerated, distorted, selected, ignored, and otherwise created a world of its own. Nevertheless, such poetry derives its themes from reality, from which it forms a new synthesis."[54] There have been attempts to analyze the synthetic political unconscious of the *Iliad*, most expertly by Peter Rose in the chapter "How Conservative is the *Iliad*?" in his 1999 study of ancient Greek literature, *Sons of the Gods, Children of Earth*.[55] His reading of the *Iliad* challenges the view of it as a conservative text that preserves traditional formulas, motifs, and narrative patterns in a representation of a static "heroic" world of continuity and homogeneity. It is true that there is an inevitability to social and political dominance in the poem going on, in tandem with "disproportionate access to all resources. Captured spoil is allotted in order of precedence. There is something of a pay scale: leaders earn the most, common soldiers earn the least . . . High-ranking Greeks are allotted the most women and the most attractive women."[56] Yet this system is unstable and vulnerable to challenge. It is also built on a paradox.

The poem, Rose therefore proposes, reveals the absolute contradiction inherent in an economic system where the expectation of orderly transmission of inheritable private wealth is frequently undermined by the social institution of warfare. Rose deftly explores the tensions inherent in the poem's attitude toward the inherited power and plutocratic values embodied in Agamemnon, and the greater meritocracy represented by Achilles. Through a subtle examination of the epic poet's own ambiguous class position, and consequent perspective on the aristocratic class, Rose suggests why the injustices (especially those suffered by women) and contradictions of "Homeric society" can be so incisively portrayed.

Rose uses the Marxist idea of "alienation" socially, to understand how, for example, Achilles becomes excluded from his group.[57] But reading the political unconscious of texts has historically not acknowledged sufficiently the effects that humans have on the environment, and their alienation from its natural processes of (re)production, any more than traditional studies of the natural world in literature have. One such traditional study is William Brockliss's *Homeric Imagery and the Natural Environment*, which, despite its aesthetically sophisticated discussion of vegetal imagery and such images' relationship with sex and metaphysics, is not concerned with relating it to environmental degradation.[58] There is also a question of whether an aesthetic appreciation of nature necessarily brings us closer to it or alienates us from it by objectification.[59]

The Emergence of Literary Ecocriticism

My own thinking about environmental issues in relation to classical literature corresponds with the historian Dipesh Chakrabarty's acknowledgment that all the insights he has gained from Marxist analysis of capital and postcolonial theory had not prepared him for making sense of the planetary crisis.[60] Donna Haraway has coined the term "naturecultures" to indicate the entanglement of environmental issues with both global politics and individual human communities' experience of their living standards and biospheres on a local level; this has helpfully been used in a fine discussion of Achilles's battle with the River Scamander by classicist Brooke Holmes.[61]

There have been pioneering studies of the relationship between much more recent literature and the environment since Henry Nash Smith's *Virgin Land* (1950), Roderick Nash's *Wilderness and the American Mind* (1967), and Raymond Williams's seminal *The Country and the City* (1973). A few projects scattered across different disciplines within the humanities in the 1980s led to the foundation of the Association for the Study of Literature and the Environment (ASLE) in 1992. The association's journal, *Interdisciplinary Studies in Literature and Environment*, was founded the following year. The most influential early publications in this field included Lawrence Buell's *The Environmental Imagination*, Kate Soper's *What Is Nature?*, and a book that is generally regarded as having founded ecocriticism: Cheryll Glotfelty and Harold Fromm's landmark *Ecocriticism Reader*.[62] Glotfelty defines the new approach as "the study of the relationship between literature and the physical environment. Just as feminist criticism examines language and literature from a gender conscious perspective, and Marxist criticism brings an awareness of modes of production and economic class to its readings of texts, ecocriticism takes an earth-centred approach to literary studies." She argues that ecocriticism, while applied to the time-honored category of "literature," is the result of something far more recent, that is, "the troubling awareness that we have reached the age of environmental limits . . . when . . . human actions are damaging the planet's basic life support system. We are there. Either we change our ways or we face global catastrophe, destroying much beauty and exterminating countless fellow species in our headlong race to apocalypse." Ecocritical practice has indeed now asked how Marxist and neo-Marxist models of the political unconscious can be harnessed to the environmental cause.

Lee Medovoi, professor of English at the University of Arizona, has coined the new term "biopolitical unconscious," marrying Jameson's concepts to what Michel Foucault in the late 1970s labeled "biopolitics."[63] This term is used to characterize the point in the early nineteenth century, when, Foucault suggests, life itself first became a political issue, a development connected with the massive population growth and demographic changes caused by the industrial revolution. Life—human, animal, and plant—the Greek *bios*, became central to capitalism's mode of regulation and to the geopolitical struggle over the power circuits between organic life on Earth and material accumulation. Medovoi explains ecocriticism, the search for what he calls the "biopolitical unconscious" of texts, in these lucid terms:

> To interpret texts ecocritically is to read them in relation to the run-up to a human-generated eco-catastrophe that threatens, not exactly the planet itself, but the "biosphere," planetary life in all its human and nonhuman forms. For this reason, ecocriticism often takes itself to be both fully historical in its perspective and planetary in the scope of its concern. Its readings of literature in this sense work backward from the crisis-ridden present (either openly or tacitly) to the origins and development of either the human

attitudes and practices that have led to the brink of such disaster, or else to alternative human attitudes or practices that might help us to avert it.[64]

This brings us back to the concept of anthropogenic damage to the environment and its deep history. Timothy Clark, in *Ecocriticism on the Edge*, notes that, for all its problems, the term "Anthropocene" can usefully serve as a "threshold concept," prompting new ways of thinking simultaneously about both short-term developments and deep-historical processes.[65] Almo Farina has proposed a transdisciplinary new model of (rather imprecisely defined) "ecosemiotics" as a unifying epistemological tool that could help all academics in all scientific and humanities subjects understand better the relationship between human and natural processes; it could offer an intellectual paradigm applicable to all organisms (humans, animals, plants, fungi) that perform procedures to stay alive.[66] Others have attempted to ground ecocriticism in the work of philosophers notorious for their obscure modes of expression, especially Heidegger, not always successfully.[67]

More promising in the case of interpreting the *Iliad* is classicist Christopher Schliephake's view that ecocriticism, when applied to canonical literature, is just one strand in the much broader project of environmental humanities.[68] In his seminal 2020 book *The Environmental Humanities and the Ancient World*, Schliephake argued that the ancient Mediterranean world represents a crucial chapter in the deep-historical process, one that needs to be investigated: "antiquity makes up the deep perspective of historical consciousness . . . the popular narrative of the Anthropocene . . . makes use of the generic structure of rise-and-fall narratives inspired by ancient models and elaborated in (early) modern historiography." This is because, "culturally speaking, an Anthropocene existed long before its geological and material effects became apparent." He points to the pronounced ecological aspects of the *Epic of Gilgamesh* and the Cyclops of the *Odyssey*; they live in harmony with the environment but can't make full use of its abundant riches.[69]

Schliephake himself uses for his extended case study the religion of classical Athens, which he uses to illustrate the dialectic of an "ecologised history" and a "storied ecology," drawing on Greg Garrard's call for the dual project of historicizing ecology and ecologizing history: "all history is environmental."[70] But Schliephake also laments the relative lateness of classical scholarship to ecocriticism, and especially the frequent absence of any ancient Greek or Roman material in anthologies or volumes of collected essays on ecocriticism or environmental humanities.[71] He argues:

The time is ripe for an "ecological turn" in classical studies/ancient history . . . Many of the motifs and symbols of the classical canon are very useful in this context because they are generally well known and carry their own weight as cultural sediments in a layered natural-cultural history . . . They are but small parts in the vast, vibrant mosaic we now refer to as the environmental humanities.[72]

Schliephake has also edited an invaluable collection of essays that further his aim of uncovering "the ancient discursive modes of literary ecology," with particularly suggestive treatments of Lucretius, Statius, and the ancient pastoral.[73] More recently, the collection he co-edited with Evi Zemanek, *Anticipatory Environmental (Hi)Stories from Antiquity to the Anthropocene*, aims to examine "the interplay of environmental perception" with "the way societies have thought about the future state of 'nature,'" by "looking at the way they have narrated and mediated this form of environmental imagination."[74] The essays include significant ecocritical studies of Hesiod, Lucretius, Galen, Plutarch, and the Roman wetlands.

Ecocriticism and World Literatures

Timber was squandered across the Ancient Near East. The economic records of Mari (in the Euphrates in Syria) in the early second millennium BCE show a vast demand for wood for smelting, chariot making, and building.[75] In the eighth century BCE, Sargon led his army against the king of Urartu in the Zagros mountains. These "high mountains" were "covered with all kinds of trees, whose surface was a jungle, over whose area shadows stretched as in a cedar forest . . . Great cypress beams from the roof of his substantial palace I tore out and carried to Assyria . . . the trunks of all those trees which I had cut down I gathered them together and burnt them with fire."[76]

The essays in Parham and Westling's collection on global literature and the environment demonstrate how much the field of ecocritical classics has to learn from analyses of other cultures. Stephanie Dalley has shown how the literature of the Sumerians, followed by that of the Babylonians and then the Assyrians, reflected their inhabitation of flat alluvial plains, marshlands, mountains, and steppes. The Sumerians created a world of myth in which nonhuman organisms are endowed with speech in a sustained manner that marks the literature off from the world of the *Iliad:* animals can speak like humans. In the Babylonian legend of Etana, the palm tree and tamarisk engage in a debating contest.[77] But, just as in

Homeric epic, abundant and almost infinitely variable similes and metaphors are drawn from the countryside—people assemble like flocks of birds, and kings describe themselves as wild bulls, hunters, raging flood waves, and lions; it was the king's prerogative to kill lions.[78]

The Mesopotamians' literature displays their deep consciousness of their dependence on the great rivers; they buried their kings in positions that are thought to have been selected to keep the river flowing in a direction helpful for humans. The coffin of a giant hero called Orontes was discovered when an unknown Roman emperor diverted the river.[79] Their literature also reveals an awareness of the sustained deforestation of the mountains, which was undertaken to ensure sufficient timber imports across the centuries. Mining and processing of metal ores "certainly created poisonous conditions along with deforestation for fuelling furnaces. Salination and erosion in fertile alluvium are also alluded to in the mythology."[80]

Interactions with nature by both gods and humans have both beneficial and destructive results. In the Sumerian *Lament for Sumer and Ur,* the god Enki could alter the course of the two great rivers Tigris and Euphrates, and make bad weeds grow on their banks, "banishing hoeing, seeding, the tunes of cowherds' songs, ensuring no more butter and cheese made in the cattle-pen, no more dung [for fuel] stacked on the ground, the marshes so dry as to be full of cracks."[81] The god Ninurta helpfully used stones to divert mountain rivers into the Tigris, but he drowned humans by directing ferocious torrents against them, and sent sudden floods that drove boulders downhill to crash into hapless travelers.[82] The gods also acted to protect nature from human depredations. The *Epic of Atrahasis* reports divine attempts to control human overpopulation by regular instances of famine, flood, and disease; Enkidu has to die because he sacrilegiously cut down trees.[83]

Although there have been occasional attempts to identify environmentally friendly strains in the biblical creation story, most ecocritics follow Lynn T. White's seminal 1967 essay "The Historical Roots of Our Ecological Crisis."[84] White pointed to God's instructions in Genesis of the Hebrew Bible as an ominous license for humans to bend nature to their will: "And God blessed them, and God said unto them, Be fruitful, and multiply, and replenish the earth, and subdue it: and have dominion over the fish of the sea, and over the fowl of the air, and over every living thing that moveth upon the earth." Multiply, subdue, have dominion. There is no such divine charter for human domination of other fauna in the *Iliad,* but the behavior of its warriors can be seen as such injunctions being carried out in practice.[85]

The mindset revealed in early Chinese literature is not dissimilar. There is certainly a factual awareness that humans were refashioning the landscape by removing trees, terracing, diking, building dams, and diverting. But portraits of landscapes affected deleteriously by human agency are "greatly outnumbered" by images of natural abundance.[86] There is even *praise* for deforestation in the *Classic of Poetry* (1045–221 BCE). It is not until the much later Tang poet Liu Zongyuan's "Xing lu nan" (Troubles on the road, circa 800 CE) that we find a condemnation of profligate squandering of natural resources and a description of officials in charge of forest management who have allowed axes to cut down trees on a thousand hills.[87]

Murali Sivaramakrishnan notes similar contradictions in ancient Indian literature. The ostensible spirituality of the *Upanishads*, which celebrate a whole, collective nature, and Hindu ritual performances that encourage humans to enter into symbiosis with the natural world, are wholesale evasions of the consistent and continuous damage caused by human interference.[88] Far more emphasis was put on humans' responsibility to sustain the natural renewal of the environment, especially of forests and maize crops, by ancient Maya thought.[89] In reference to Islamic texts, meanwhile, Sarra Tlili argues that, although the Muslim theme of the worship of God by creation has been supposed by some to have environmental potential, historically speaking, the notions of stewardship and of unity of being have been invoked to testify not to humans' responsibilities toward other creatures but to humans' privileged status.[90]

Christopher Abram's exciting and flamboyant recent study of ecological catastrophe in Old Norse myth and literature traces their preoccupation with the idea of an oncoming catastrophe, Ragnarök, when the entirety of creation will be destroyed and the conditions that make organic life possible eradicated. He makes a persuasive case that this scenario was the product of an age in which Scandinavia's forests had been depleted and early Icelanders were struggling to make a viable life in a profoundly inhospitable environment.[91] Medieval European literature is always fascinated with forests, and usually ambivalent about them, for example the sinister but fertile forest in Malory's *Morte d'Arthur*.[92]

Ancient Greek and Roman literature has of course received some ecocritical attention, besides Holmes's pioneering article mentioned earlier.[93] Chris Eckerman's interpretation of Pindar's *Olympian* 7 draws attention to the interwoven naturalization of male violence against women and human violence toward the environment in the narrative of the original creation of the island of Rhodes.[94] Elizabeth Heckendorn Cook provocatively suggests that Ovid's accounts of trees that can move and transmute into humans (or vice versa) are connected with the actual mobility of trees in that era: as forests receded and shrank up mountainsides, logs were transported around the Roman Empire, and ships made from those logs sailed around the Mediterranean and Black Seas.[95] But the preeminence of Homer in the history of ancient and more modern literature makes the need for an ecocritical study pressing. Homeric poetry has profoundly informed subsequent literature and society in many and diverse ways, from providing exemplars of human heroism, love, death, and conflict as well as representations of speech and action, to creating expectations of tone, image, grandeur, scale, and magniloquence.

The Scepter Shorn

The first significant object to be described in detail in the *Iliad* is the ceremonial staff, studded with golden nails (1.246), which was held by the man addressing the assembled Achaeans. It is not, apparently, Agamemnon's own regal scepter that Hephaestus "forged with toil" for Zeus (*kame teuchōn*, 2.101), and that has come down a line from Hermes, Pelops, Atreus, and Thyestes to Agamemnon (2.102–7).[96] The studded staff is instead wielded by Agamemnon's opponent in the great quarrel that opens the epic. Achilles himself describes its wooden core as he swears the oath that will keep him off the battlefield for the first two-thirds of the poem (1.230–39):

> People-devouring king!—for it is nobodies you rule over,
> since otherwise this would be your last act of outrage—
> I, however, will speak out to you, and swear a mighty oath to support
>   it:
> By this staff, which has not, and never will, put forth leaves and shoots,
> since first it left its stump behind in the mountains,
> and which will never sprout again, for the bronze
> has shorn it of leaves and bark, and now the sons of the Achaeans
> hold it in their hands when they are dispensing justice,
> the men who guard the ordinances that come from Zeus.

Agamemnon has trampled on the divine ordinances, the taboos and imperatives laid down by Zeus, in failing to respect his greatest warrior. Achilles implies that this is not the first time that Agamemnon has committed an outrage, and that he has damaged his entire people in doing so.

But the idea of damage is visually expressed in Achilles's strange description of the scepter—it was once alive and vibrant but has been rendered lifeless and infertile by the application of human technology.

I hear this description differently from most scholars, who, following Nagy, have seen the tree's transformation into a cultural object as somehow positive, something to be celebrated as a human achievement. Nagy believes it is the same scepter as Agamemnon's own, which is elsewhere said to be "ever unwithering" (*aphthiton*, 2.46). This term implies "the cultural negation of a natural process," characterizing both Hephaestus's creations (14.238–39, 18.369–71), and epic poetry's function. According to Nagy, Achilles achieves unwithering renown (9.413) "as compensation for the death that he cannot escape."[97] But neither manufactured objects nor epic poems are unwithering. As mentioned earlier, some experts believe that if we do not mend our ways, we have only about five hundred years before the planet will undergo the next Great Extinction.[98] Classics, and the *Iliad*, are not even three millennia old.

An alternative way of hearing Achilles's description of the scepter that the Homeric imagination has produced is that it—implicitly, at least— likens the crimes that an unjust monarch inflicts on his people, both aristocrats and the *dēmos*, to the anthropogenic violence done to a tree by cutting it down at the stump and stripping it with the sharp products of metallurgy. This detailed description of an artifact, the first such description in the poem, thus sets the agenda for the remaining twenty-three books' exploration of the relationship between humans' exploitation of other humans and their exploitation of the organic world. It is the master symbol of the *Iliad*'s political ecology.

CHAPTER FOUR

Nature and the Divine

THE *ILIAD* PORTRAYS A world in which humans thought to have existed several centuries before the poem's original audience struggle to understand the workings of their surroundings—the environment in which they live. This environment includes flora, fauna, minerals, earth, rivers and seas, cosmic entities such as the sun, stars, planets, and meteorites, and meteorological and elemental forces including wind, rain, snow, clouds, rainbows, and fire. These phenomena are sometimes portrayed as acting independently and autonomously, but often as being harnessed by one of the anthropomorphic gods, notably the thunder and lightning sent by Zeus. Hephaestus uses fire in his foundry and can drive wildfire across the Trojan plain. But on three occasions his name seems nearly or actually interchangeable with that element. Hector cries a piercing cry "like the unquenchable flame of Hephaestus" (17.87–88); many animals "were stretched and singed over the flame of Hephaestus." The Argives even cook meat on spits "over Hephaestus," where the metonymy is complete.[1]

The word "metaphysics" is now commonly understood, as it was by some medieval commentators on Aristotle, to mean "the science of what is beyond the physical."[2] Two strands that are metaphysical in this sense are shared by the fictional world conjured in the *Iliad* and other early Greek thought: both anthropomorphic gods and elemental principles are divine, but some physical principles are more sentient and forceful than others. The boundaries between anthropomorphically conceived deities and physical, elemental, and meteorological phenomena are unstable; the River Scamander can suddenly assume the likeness of a man and speak to Achilles (21.13). There are scant traces of Hesiod's schematic presentation of natural phenomena such as Night and Earth as arising prior to the arrival of the Olympian gods, although Ocean seems to possess a special status as a god from whom others originated.[3] As Peter Karavites explains, Homeric poetry shares with Ancient Near Eastern poetry no assumption of "any absolute separation between the animate and inanimate worlds on the one hand, or between the divine and humans on the other . . . in the Homeric epics the obvious remnants of animism and totemism diminish if not eliminate the distance between man and nature, whereas anthropological gods are organized in societies and lineages resembling human societies and lineages."[4]

Some natural phenomena are acted on or used by humans, especially as they hunt, log, farm, and work metal. Conversely, natural phenomena frequently affect humans in ways that are beyond their control. A few of these phenomena are semi-personified by gods. Iris is in a category of her own. A meteorological rainbow is called an *iris* in a context that does not mention gods (11.27). No doubt the idea of messages and messengers from the gods was an intuitive explanation of the colorful bands of sunlight refracted through raindrops that suddenly bridge sky and land or sea before evanescing.

But Iris herself is never described as looking like a rainbow while she delivers messages for Zeus and Hera (and on one occasion Achilles, 23.199), and sometimes she acts autonomously (3.122, 15.201, 18.197). The swiftness of rainbows' manifestation in the wake of rain is suggested by her epithet "storm-footed" (8.409; 24.170) and the comparison of her when she is in motion to snow or hail driven by wind down from the clouds (15.170–71). Priam recognizes her face but does not describe her appearance (24.223). Iris seems to be imagined as an anthropomorphic being with golden wings (8.438), rather than an envoiced meteorological rainbow. She is sent by Zeus to Hector, and stands by him as she tells him to hold back until Agamemnon leaves the battlefield (11.199–209). Hector clearly understands from this that Zeus has granted him glory, but we are not told if he can see Iris (11.284–91). Hera sends Iris to Achilles to instruct him to enter the fray; even the other gods do not know. Achilles recognizes and converses with the rainbow goddess but, again, does not describe her (18.166–200). A rainbow can also represent a quite different divinity. Athena first covers herself in a purple cloud (17.551–52), then disguises herself as Phoenix to speak to the Achaeans; this makes an impact similar to the

purple rainbow that Zeus sends as a portent, which makes men stop working and upsets their flocks.[5]

Rather different semi-personifications are Dawn and Helios, whose repeated cosmic movements are events that mark the passing of time; night,
despite being "immortal" (*ambrosiēn*, 9.41), is not quite in the same category. Dawn, with her fingers like roses (1.477), is an embodied being, a
goddess (2.49), imagined rising from her couch beside Tithonus to bring
light to gods and men (11.1–2); she wears a saffron robe as she spreads
across the Earth (8.1, 23.227, 24.695). Yet the boundary between corporeality and being constituted by a kind of sentient force field here seems
porous. Even *partial* embodiment is less clear in the case of Helios, although he can be the recipient of sacrifice at an oath-making ritual (3.104),
where Agamemnon prays to him in the second person singular (3.277) as
"you who see all and hear all." Helios has emotions: when Hera sends him
prematurely to the streams of Ocean, he goes "unwillingly" to set there
(18.239–40). Embodiment perhaps seems implied when he is named Hy-
perion (8.480, 19.398), the name of a Titan to whom Gaia gave birth. Gaia
herself receives the oath sacrifice at the same time as Helios and the Rivers
(3.278). She, meanwhile, is envisaged as groaning when many people are
assembled in a single place (2.95, 2.785), and laughing in delight when
Achilles arms himself to return to the fray. But at other times earth is an
apparently inanimate material element, like water (7.99, 23.256).

Both Gaia and probably Helios are already present in Linear B, although the Minoan-Mycenaean deity with a solar aspect seems to have
been the female great goddess.[6] Yet one of the overall continuities the
poem assumes between the Mycenaean-era world portrayed in the *Iliad*
and the world of the eighth to seventh centuries BCE that produced the
poem was the worship of approximately the same group of gods, often with
similar portfolios, such as Poseidon's connection with the sea. Although
there was huge variety in emphasis on different gods and ritual practices
across different Mycenaean communities, the pantheon that has emerged
from the deciphered texts of Linear B suggests that this assumption of
some continuity by the epic poets was by and large correct.[7] Although
Apollo and Aphrodite have not been identified in the Mycenaeans' writings, and may have entered the Greek pantheon from Near Eastern reli-
gions slightly later, most of the other gods named in the Homeric poems,
including Dionysus, are already there in Mycenaean Greek and are already
identifiable as precursors in terms of their spheres of competence.

An attempt to locate and describe the ecological unconscious of the
poem, therefore, must ask how Homer's original audiences, who had in-

herited and substantially developed Mycenaean religion and Mycenaean poetry over several centuries, understood the involvement of the gods in the natural world with which both they and their Iliadic heroes interacted. Humans are portrayed as ruthless exploiters of natural phenomena, most obviously in their profligate consumption of timber, livestock, and metal. But they are also portrayed as acknowledging their own precarity in the face of extreme weather and freak forest fires; they are presented as vulnerable, too, to the plague in the opening book, feverish diseases (21.31), and, however briefly, to famine in the final one: Achilles tells Priam that in the urn containing evil gifts for mortals, which stands in Zeus's palace, there lies ravening appetite or famine (*boubrōstis*, 24.532), which drives a man across the face of the Earth to its very ends, respected by neither gods nor men. Much less exposure is given to the injuries that humans might inflict on natural phenomena and the environment, although, as we shall see, there are occasional flashes of prognostic insight.

The authorial voice of the poem lays claim to possessing a far greater understanding of the activities of the gods and the role they play in the Trojan War, including their influence over nature and the elements, than is enjoyed by the human *dramatis personae*. This is most obviously apparent in the accounts of interactions between gods where no human is involved, such as the divine assemblies on Olympus, Hera's instructions to Hephaestus to intervene in Achilles's battle with the River Scamander (21.328–41), or the banquet of the Winds at the house of Zephyros visited by Iris (23.198–211). Another striking example is the authorial persona's knowledge of the different words the gods use for certain phenomena, such as the mound in the Trojan plain known to men as Batieia, but to the immortals as Myrine's barrow (2.813–14), or the Trojan river that men know as Scamander but the gods call Xanthos (20.74).

Perhaps the authorial persona is enabled to access this knowledge, imperceptible to other mortals, by his special relationship with the goddesses of poetry, the Olympian Muses, daughters of Zeus. They are usually invoked in the plural to empower the persona to remember complex lists of proper names, for example the Achaean leaders in the catalogue of ships: the authorial voice asks them who these leaders were, "for you are goddesses and are present and know all things, whereas we hear only a report (*kleos*) and know nothing" (2.484–85; see also 11.218, 14.508, 15.112). But the Muses are no nebulous metonymy for "inspiration" or "memory"; they are proud, embodied beings. They have been known to hold a singing contest with a human bard, Thamyris, who boasted of his superior musical skill, lost the competition with his several victors, and was blinded by them

(2.594–600). The authorial persona, however, invokes a singular "goddess" in the poem's opening lines, whom he asks to sing of the wrath of Achilles that sent warriors to Hades, thus fulfilling the plan of Zeus (1.1–5). The goddess will sing, for/through the poet, of a cosmic system in which mortals are glorious, vulnerable, and centrally implicated. It already includes a god of the Underworld, to whom the war dead were sent, and a divinity named Zeus who was implementing a plan.

The spheres of divine, human, and even animal activity, although sharply distinguished in terms of different relationships with time through (im)perishability and (un)susceptibility to death, often intersect.[8] The boundaries between them are otherwise strikingly permeable. Zeus regrets having bestowed the ageless immortal vocal horses Xanthos and Balios on the mortal Peleus, thereby making them suffer sorrow among men when they mourn Patroclus, since men are the most wretched beings that breathe and move on the Earth (17.442–47). Xanthos marks, by disrupting, the normal strict boundary that prevents humans and animals communicating verbally when he warns Achilles, finally about to reenter battle, that the day of his death draws near. He will be slain by a god and a mortal (that is, Paris and Apollo; see 22.358–60). This instance of permeability is underlined when we are told that it was the Erinyes who stopped Xanthos from speaking, since this implies that he has transgressively destabilized the rightful boundary between man and animal, even an immortal animal, a boundary that prohibits verbal communication (19.404–19). It is interesting, too, that horses with riders, apparently associated with particularly venerated deities, appear among the terracotta figurines found in a Mycenaean sanctuary at Methana in the eastern Peloponnese.[9]

The immortal/mortal borderline is further blurred because Achilles's mortal horse Pedasos keeps pace as a trace horse beside these immortal steeds (16.148–54). Rhesus's enormous horses, whiter than snow, and as fast as winds (10.436–37), are so "terribly like the rays of the sun" that the battle-hardened Nestor thinks they must have been given to Odysseus and Diomedes by a god (10.545–51). Such conjunction and resemblance between mortal and immortal animal are paralleled, in a way, by the River Titaressus near Dodona, which, although flowing into the (today heavily polluted) River Pineios, does not mingle waters with this terrestrial river, but instead only floats like olive oil on its surface, because it is a branch of the immortal Underworld River Styx (2.749–55).[10] Similarly, mortals sometimes march into battle with gods alongside or leading them.

Gods can interact with humans everywhere that humans are to be found: most frequently on the battlefield, where Apollo returns into the fray, "a god into the toil of men," but also in tents and houses, on the road, and on hillsides where humans are herding livestock.[11] Yet mortals are excluded from some of the haunts of the gods, notably Mount Olympus, which seems almost completely inaccessible to human presence.[12] The exception here is Ganymede, whom Zeus extracted from his terrestrial existence to serve him among the immortals (20.233–35).

Mortals Like Gods and Favored by Them

The mortals portrayed in direct individual interaction with immortals are also all aristocrats, unless ordinary soldiers are to be included in the single instance where all those who fought at Troy are described collectively as the "people who were men half-divine" (*hēmitheōn genos andrōn*, 12.23). The close connections drawn between some nobles and some gods are based on consanguinity, patronage, or similarity; both the biological mixing between gods and humans and the emotional ties binding them are features shared with Ancient Near Eastern mythology.[13] Like Helen, the daughter of Zeus (3.418), some of the warriors are half-divine; Sarpedon, too, is the offspring of Zeus. With supreme tactlessness, Zeus says he feels more desire for Hera than when he lay with the mortal mothers of Peirithous, Perseus, Minos, Rhadamanthys, Dionysus, and Heracles (14.313–25). Other gods besides Zeus have children fighting at Troy, further complicating the networks of allegiances between the immortal and mortal spheres; the Achaean Ascalaphus is a son of Ares (13.518–26), Aeneas's mother is Aphrodite (2.820), and the father of the Myrmidon Eudorus is Hermes (16.179–86).

Divine parentage can also involve a natural phenomenon, usually a watery one. The mother of the two sons of Talaemenes was the nymph of the Gygaean lake (2.865), Aesepus and Pedasus were sons of the fountain-nymph Abarbaraē (6.21–22), and Iphition was son of a Naiad (20.384–85). The Myrmidon Menestheus was son of the River Spercheios, which is made of "Zeus-fallen" water, that is, rain (16.174; sadly, the volume of water in this river has been shockingly depleted as a result of human exploitation and interference).[14] The question of the status of children of river gods becomes particularly important in book 21.

Warriors with mortal fathers and immortal mothers include Achilles, whose mother is the Nereid Thetis (although he believes that this has caused him to suffer and wishes that his father had taken a mortal bride;

18.86–87). Achilles's watery entanglements also include fights with a river's son and a river in book 21, and interestingly, a close relationship with the River Spercheios himself. Achilles cuts off a lock of hair to bestow on Patroclus's funeral pyre, but reveals that he had actually grown his hair in honor of Spercheios; this assumes a cult, perhaps a kourotrophic one, of the Myrmidons' river god.[15] Achilles's father, Peleus, had vowed that upon Achilles's return from Troy, the hero would shear off and dedicate his hair to Spercheios, before sacrificing a hecatomb of cattle (an offering theoretically consisting of a hundred slaughtered animals), plus fifty unblemished rams, and putting them in his waters near his altar (23.141–48). Hesiod's injunctions include a ban on urinating or defecating in river or springs; this horror of soiling supplies of fresh water perhaps does not sit well with the information that the Trojans used to sacrifice bulls to the River Scamander and cast living horses to their deaths in his streams (21.131–32).[16] The sacred qualities of rivers are suggested in the *Iliad* when Zeus decrees that the body of his beloved son Sarpedon must be washed in the streams of the river (16.669) before being anointed with ambrosia and clad in immortal garb.

Other heroes are a quarter divine, in that one parent was a demigod, such as Tlepolemus, famed for his spear, born to mighty Heracles by Astyocheia (2.658), or Polypoetes, whose paternal grandfather is Zeus (2.740–42). The death and beheading of his grandson Amphimachus make Poseidon incandescent with rage (13.206–7). Heroes who are one-eighth divine include Idomeneus, paternal great-grandson of Zeus (13.450), as well as Crethon and Orsilochus, paternal great-grandsons of the River Alpheus (5.541–49; this great river is today wrecked along its entire course by excessive damming, diversion for irrigation, gravel mining, and absorption of human waste and chemical fertilizers).[17]

Homer imagines that a mortal's half-divine parentage can be used by a god to argue, to another god, that the hero should receive special favor. Hera protests to Apollo (24.58–61):

> Hector is a mortal and suckled at a woman's breast,
> but Achilles is the offspring of a goddess whom I myself fostered
> and raised and bestowed on Peleus as his wife,
> a man extremely dear to the gods.

But what if two rivals are both half-divine? Then each one's status relative to another one's may depend on the relative eminence of the parent within the divine hierarchy.

Apollo, disguised as Priam's son Lycaeon, encourages Aeneas to fight Achilles on the ground that Aeneas's mother is Aphrodite, a daughter of Zeus, whereas Achilles's mother is a lesser goddess, daughter of the old man of the sea (20.104–7; see also 20.206–9). It is important to note, however, that there is a sense in which being *honored* by a god, especially Zeus, can trump being *born* of a lesser deity. As Nestor says to Achilles, he should check his rage, since although he is a stronger fighter and born of a goddess, Zeus has awarded Agamemnon a special honor and glory by bestowing upon him a kingly scepter (1.276–81). Zeus observes that Hera favors the Achaeans so fiercely that they might as well be children of her own womb (18.357–59). And even some warriors with a more remote ancestral link to the immortals are described by epithets implying their affinity or connection with them. Gods, heralds, and individual humans can be described as "dear to Zeus."[18]

A woman can be "like the goddesses."[19] Women are said to contend with Aphrodite in beauty and Athena in handicrafts (9.389–90). The simple term "godlike" (*dios*) embraces a large cross-section of immortal, natural, and mortal entities. It is used to refer to goddesses (10.290, 1.388. 19.6), and to Artemis as "divine offspring" (9.538); perhaps this implies that male gods are already so *dios* that it would be tautologous to say so. A numinous divinity, too, seems to lurk behind some grammatically feminine but not explicitly personified natural phenomena such as the sky, sea, and land, since they are also lent this epithet.[20] Adrastus's horse Arion, said in later sources to be the offspring of Demeter and Poseidon when they mated in equine form, is described as *dios* (23.346).[21] But Hector addresses Lampos, one of his team of horses that are *not* explicitly said to be sprung from gods, in the vocative as "*die*" (8.185). Among mortals, *dios* applies to men as well as women (Achilles, born of a goddess, but also, for example, Agamemnon, 2.221; Odysseus, 9.169; Hector, 11.197), and to collectives such as the Achaeans, Sarpedon's loyal comrades, or a city such as Elis (5.451, 20.354, 5.692, 2.615). Divinity, or at least the affinity with a male god that the epithet *dios* implies, is something in which female gods, natural entities (feminine sea, sky, land, a male horse), and mortals of both genders can partake.

There is a further range of epithets and formulae for male heroes that stress an equivalence to or affinity with gods: "equal to the gods" (*isotheos*, for example of Euryalus, 2.565), "godlike" (*antitheos*, Sarpedon, 5.633; the Lycians, 12.408), "like the immortals" (*epieikelon athanatoisin*, of Acamas, 11.60), "godlike in form" (*theoeidēs*, most notably of Priam at 24.217), "nurtured by Zeus" (*diotrephēs*, 2.196); "many cities of lusty warriors nurtured by Zeus" (2.660, 17.679) and "sprung from Zeus" (*diogenēs*, of Odysseus,

4.358). Apollo fights hand-to-hand with Patroclus, three times thrusting him back "with his immortal hands" from the wall (16.702–4); at the very point when Patroclus, fully mortal but undaunted by fighting with the god, is said to be "equal to a divinity" (*daimoni isos*, 16.705) as he makes a fourth attempt, he is addressed by Apollo, who puts a stop to Patroclus's progress, as "sprung from Zeus" (*diogenēs*, 16.706).

Zeus himself, the supreme deity, is never compared with anything or anyone else, having inherited from Ancient Near Eastern literature a legacy of incomparability.[22] But Hector is "the peer of Zeus in intelligence" (7.47, 11.200); both he and Meriones are peers of "speedy Ares" (8.215–16, 13.528). Nestor addresses the Achaeans as "servants of Ares" (2.110, 6.67). Warriors including Elephenor and Leonteus are described as an "offshoot of Ares" (2.540, 12.88). "Dear to Ares" is used of Achilles, Menelaus, and the Achaeans (for example, 3.21); Menelaus, Ajax, the "sons of the Achaeans," weapons, arms, and even a wall of soldiers can be described as "belonging to Ares" (4.407, 15.736). In these cases, the stem "Are-" is synonymous with "war," and the epithet can be translated as "martial." Where formulae and epithets involving Ares are concerned, it is indeed often difficult to tell whether the proper name is simply a metonym for an impersonal, abstract concept of "war" or "armed combat": "sharp Ares" has shed the blood of many Achaeans (7.328–30); the poet speaks of "the toil of Ares" (16.245); Trojans and Achaeans share in "the fury of Ares" (18.264); "Ares destroyed them" (23.260).[23]

On one spine-tingling occasion, we hear that Ares enters a hero physically. When Zeus makes Achilles's armor fit snugly around Hector (17.209–10),

> Ares entered him,[24]
> dreaded Enyalios [god of the war-cry], and his limbs were filled
> with valour and strength.

The addition of the epithets, especially Enyalios, with the reference to his vocality, makes it difficult to read Ares in this instance as a mere metonym for the abstract idea of a warlike mood. Similes also liken warriors to Ares: Ajax storms into battle "like enormous Ares when he joins battle amidst men" (7.208–9); Meriones and Idomeneus go on the attack like Ares, "plague of men," accompanied by his henchman Phobos (Terror), marching out from Thrace (13.298–302).[25]

The similarity between gods and men is stressed in other distinctive passages of rhetoric. Apollo may tell Diomedes to desist from fighting gods, "since the tribe (*phulon*) of immortal gods and that of humans who walk the earth are in no way the same" (3.441–42), but Achilles says he will go against the Trojans despite their huge numbers, which would be impossible even for an immortal god like Ares or Athena to control (20.368–69). Priam says that Hector was "a god among men," who seemed as if he was son of a god and not of a mortal man (24.258–59).[26]

Finally, some mortals enjoy special gifts that the gods or supernatural beings have personally bestowed upon them. As Polydamas puts it, the god has given the skills of war to one person, dancing to another, music-making to a third type, and intellect to another (13.730–34). The reader of the *Iliad* is invited to notice whether characters have a special gift given to them by a god, and if so, to pay closer attention to how that gift is deployed. The kind doctor Machaon has been given effective medicinal drugs by his father, who received them from Cheiron the centaur (4.218–19), so we pay closer attention to how effective they are when he treats the wounded Menelaus. Paris's hair and physical beauty are gifts of Aphrodite, which adds glamor to the scene where he seduces an initially reluctant Helen after Aphrodite rescues him from the battlefield (3.54–55, 3.421–48). Bellerophon was granted beauty and manliness by the gods, but Zeus had given Proetus the scepter of monarchy; this helps to explain why Proetus's wife fell in love with the handsome Bellerophon despite the kingly power wielded by her husband (6.156–62).

Agamemnon's scepter was made by Hephaestus and given to his ancestor Pelops by Hermes, who had it from Zeus (2.100–109), which focuses attention on his abuse of kingly power. Other gifts help to round out the portraits of both the divine donors and the mortal recipients. Diomedes's breastplate was also made by Hephaestus (8.195), which underlines the special status of this warrior, whose performance on the battlefield in book 5 would be worthy of Achilles. "Bronze Ares" had given his armor to Areithoüs, whose name means "swift like Ares," (7.146). Apollo had given Teucer his bow (15.441), and Hector his helmet as a gift (11.353); this adds ironic pathos to the moment when Apollo deserts Hector on the battlefield in the hero's final minutes.

Similarly, when we learn that Andromache's veil was a wedding present from Aphrodite (22.469–72), at the very moment when she learns that she is a widow and casts it from her head, the veil's provenance reminds us of Hector and Andromache's happy marriage and sexual relationship. The gods gave Peleus wealth, kingship, and a goddess as a wife (24.534–37), but the audience knows that the goddess had abandoned him. The Thessalian Eumelus has two mares that Apollo had reared near the hero's hometown (2.763–67). Aeneas's horses are descendants of those that Zeus gave to Tros in return for Ganymede, for Anchises had secretly put his mares to them (5.260–69); this foreshadows how later several gods took special care of Aeneas, knowing that he was destined to become a Trojan king who would perpetuate the people's genetic line into the future (20.300–308). The Trojan Ilioneus is son of Phorbas, whom Hermes loved so much that he made him rich in flocks (14.489–91).

Mortals Aware of Interactions with Gods

What can mortal heroes expect to experience in terms of direct interactions with gods? Gods can visit humans without disguise and be seen and/or heard by them. Hector, sitting winded on the battlefield, converses face-to-face with Apollo (15.243–62), who puts fresh courage into him, enabling him to terrify the Achaeans again. When Apollo then leads Hector and the Trojans, he is apparently visible not just to Hector but to everyone, although his shoulders are wrapped in cloud; when he stares the Achaeans full in their faces, shakes the aegis and shouts, they are all petrified (15.306–22).

The most elite warriors often recognize a god when he or she appears. Achilles recognizes Athena when she grabs him by the hair to prevent him from assaulting Agamemnon, but nobody else can see her (1.196–99). Achilles can summon his mother from the sea, and she appears to him physically (1.359). Odysseus recognizes Athena's voice when she comes down to the battlefield to speak to him (2.182). Mortals can sometimes even see through the disguise of a god. Aphrodite assumes the disguise of an old woman who used to card wool for Helen in Sparta, but Helen knows it is the goddess (3.385–420). Apollo disguises himself as the herald Periphas to rouse Aeneas, but Aeneas recognizes the god's face anyway (17.322–34). Diomedes injures Aphrodite and fights fearlessly even with Apollo, knowing well who he is (5.432–37). Diomedes also sees through Ares's disguise as the Thracian Acamas (5.460–71). But Diomedes has fundamentally challenged the separation of men and gods by wounding Aphrodite and fighting Apollo, perfectly aware of who they are.

There were other occasions when Olympians suffered severe pain at the hands of men (15.383–84).[27] Ares had been bound inside a bronze jar by the son of Aloeus, and Heracles had injured both Hera and Hades

(5.385–400). The last two examples are questionable, since the Aloadae are in other sources not humans but giants, and Heracles is hardly a typical mortal (he is semi-divine by birth, then divine by election).[28] Yet it is clear that gods are not wholly immune to being harmed by mortals, even though it is acknowledged that some gods, especially the warlike ones such as Ares (5.604–6), are far more difficult to contend with than others. Laomedon once hired Apollo and Poseidon to work for him, but defrauded them and threatened them with shackles, slavery, and mutilation (21.441–57). This experience convinced them that mortals are not worth fighting over, since, as Apollo puts it, they are pathetic, perishable creatures; they resemble leaves, because at one moment they "are full of blazing life" and at the next they wither and become lifeless (21.464–66). In this beautiful simile, humans in their perishability are presented as similar both to the element of fire and to vegetal matter.

A god may communicate by sending an audible omen: Athena, to show her favor, sends Odysseus and Diomedes on the right of their path a heron, a bird that, because it is night, they can only hear and not see. Odysseus asks Athena to aid their expedition on the ground that he senses that she favors him. She is "always present" during his ordeals, and she perceives every move he makes (10.274–80). Athena can breathe strength into Diomedes so that he kills twelve Thracian warriors (10.482) and through her cunning makes Diomedes appear to Rhesus like a figure of nightmare (10.497).[29]

Athena's heron is audible, but she apparently does not speak herself on this occasion. Hera draws a distinction between the impact on a mortal of hearing the voice of a god and laying eyes on one, "for the gods are difficult to look on when they appear in manifest presence" (20.131).[30] When Hector is terrified by hearing Apollo's voice warning him to hold off on fighting Achilles, it is the content of the god's words rather than the sound of his voice that causes the fear (20.375–80). "Storm-footed" Iris does not go in disguise to Priam, and he is frightened (24.170). He subsequently tells Hecuba (24.223) that he did look on Iris's face as well as hear her voice.[31]

Other naturally occurring phenomena besides the plague blighting the Achaeans (1.9) and Athena's audible heron (10.274–75) can be correctly diagnosed by mortals as a sign of divine engagement in human affairs. Andromache knows that the elm trees that grew up round her father Eëtion's funeral barrow were planted by the mountain nymphs, daughters of Zeus (6.419–20). When Hera has made Agamemnon exhort the Achaeans to action, and he has prayed to Zeus, reminding him of his frequent sacrifices

Zeus sends his eagle of omen to Priam. Drawing by Becky Brewis of
William Hamilton's painting *Priam's Prayer on his Departure for the
Grecian Camp.* (Courtesy Becky Brewis)

to him, Zeus sends the reassuring omen of an eagle grasping a fawn and
then dropping it beside Zeus's altar (8.245–50). Phoenix knows that the
wild boar that had once ravaged the Calydonians' blossoming orchards
had been sent by Artemis (9.533–42).When Priam, on Hecuba's advice,
prays to Zeus to send an eagle from the right to show that he approves of
Priam's plan to go to the Achaean camp (24.290–313), the authorial voice
tells us explicitly that Zeus heard and sent the bird-omen as requested
(24.314–20).

But on one earlier occasion, after Ajax has predicted that the day will
soon come when Hector prays to Zeus as he flees the Achaeans, an eagle
flying in from the right is interpreted by *both* Achaeans and Hector as favorable to their own side (13.821–37). The authorial voice makes no comment. It is not even made explicit that the bird has been sent by Zeus in
the first place.

Zeus communicates far more often via meteorological signs than via eagles. Some of Zeus's epithets reiteratively confirm his power over the weather, especially rain, other precipitation, and thunderstorms. They include "of the black clouds" (1.397, 15.46, 21.520), "cloud-gatherer" (1.511), "with vivid lightning" (19.121), "delighting in thunder" (1.419), "loud-thunder-clapping" (5.672), "loud-thundering" (13.624), "high-thundering" (1.354, 12.68), and "marshaller of the thunderheads" (16.298).[32] But while thunder and lightning are read by humans as visible signs of Zeus's opinions, they can also suggest that he is physically shaking the aegis at the same time (17.591–96):

> Then the son of Cronus took his tasselled aegis,
> brightly gleaming, and covered Ida over with clouds,
> and emitted mighty thunder and lightning as he shook the aegis,
> giving victory to the Trojans and putting the Achaeans to rout.

On just one occasion it is perhaps implied that lightning is a sign of Poseidon's presence. As he leaps into the front ranks of the Achaeans and shouts aloud to rally them, promising personally to lead them against the Trojans (14.374), he carries a terrible sword with a long edge "like lightning" that men are too terrified to clash with (14.384–87).[33] But usually a flash of lightning or a thunderclap is explicitly identified as a communication from Zeus.

Because the Achaeans failed to offer hecatombs to the gods after building their defensive wall, Zeus thunders throughout their feast until they all pour him a libation (7.478–83). When in Zeus's scales the Trojan pan rises, Zeus thunders aloud from Ida, but he sends the lightning flash amid the Achaeans, terrifying them (8.76–78). Zeus then prevents Diomedes from progressing further against the Trojans by thundering and sending his lightning bolt, complete with the stink of sulphur, to crash in front of Diomedes's horses and terrify them: Nestor knows full well it is a reprimand from Zeus (8.133–44). Diomedes wonders three times whether to fight Hector face to face, but Zeus thunders back three times from Ida, to show his support for the Trojans (8.169–71). No wonder Odysseus complains to Achilles that Zeus keeps showing his favor to the Trojans by emitting lightning flashes on the right (9.236–37).

But Zeus's thunder can be ambiguous. The authorial voice tells us that Zeus uses it when he hears Nestor's prayer that the Achaeans should survive (15.370–80). We are not told, however, whether the thunder is in assent or merely acknowledgment. And it is the Trojans who, not at all daunted when they hear the thunder, leap even more powerfully onto the Argives. For over 350 lines, until the end of the book, the battle at the ships remains evenly balanced, with Hector and Ajax still unable to make a breakthrough against the other. This is expressed in an unusual simile (15.410–13):

> Just as a carpenter's ruler makes a ship's plank straight,
> in the hands of a clever joiner, who is highly skilled
> in all kinds of craft, prompted by Athena,
> so evenly strained was their fighting and conflict.[34]

Zeus's thunder on this occasion did not mean that either side was about to be awarded victory, at least not yet.

Occasionally, weather portents signify more than arbitrary favor to one side or another. The terrifying phenomenon of blood-rain is sent to the battlefield twice by Zeus, once to signal that many deaths were about to occur (11.53–54); on another occasion he sheds blood-rain in honor of his son Sarpedon, who is soon to die (16.459–61). In two similes, a destructive natural phenomenon is presented as a dimension of theodicy, a direct providential punishment for human misdemeanor. Achilles is causing problems for the Trojans as smoke rises from a city burning because the gods are angry (21.522–24). In William Langland's fourteenth-century poem *Piers Plowman*, famine and social unrest are the consequences of humankind's abuse of animals and plants.[35] Similarly, social misdemeanors in the *Iliad*, are, in one passage that evokes Levantine deluge narratives, punished by bad weather.[36] Hector's groaning mares drive him ever onward (16.384–93):

> It was as when the entire dark earth is oppressed by a tempest,
> on a day at harvest-time, when Zeus pours down rain most
>   aggressively,
> when he is aggrieved and angry at men,
> who through violence give crooked judgements in the town square,
> and drive justice out, disregarding the gods;
> all these men's rivers flow in full flood,
> torrents carve furrows on numerous hillsides,
> roaring mightily as they flow headlong from the mountains
> into the purple sea, wrecking the products of human labour.

Janko associates this simile, featuring the loss of fertile cultivated terraces, with the deltaic fill and swampland in southern Laconia "caused by erosion resulting, at least in part, from human activity," at least by the time of the composition of the *Iliad;* a town in Menelaus's kingdom is indeed named "Helos," meaning "marsh," or "swamp" (2.584).[37] In the *Iliad,* physical disasters "naturally" have a metaphysical aspect when they are sent by a vindictive god.

Human Experts in Divine Signs

Working out the will of the gods can be made slightly less challenging by consulting humans who specialize in divining the immortals' intentions through perceptible signs in the physical world. Achilles prays to Zeus specifically in his capacity as recipient of a cult at Dodona in northern Greece, where the god's interpreters reside with unwashed feet (16.233–35). Yet there is always the possibility that such experts might make a false diagnosis; Priam says to Hecuba that he might not have credited a seer who divines from sacrifices, or a priest, if they had instructed him to go to the Achaeans' camp, since the instructions given by such men can be deemed false and not followed; the instruction had, however, come directly, without the mediation of a seer, from a divinity he had actually seen and heard (24.220–24).[38]

The difficulty in knowing whether to believe an interpretation of an omen is most clearly brought out in a crucial interchange between Hector and his prudent lieutenant Polydamas. The context of the interchange is that the Trojans have become reluctant to press on against the Achaean wall because an eagle has flown past on the left, carrying in its talons a writhing snake that fights back until the eagle drops it in the middle of the Trojans and flies away noisily on the paths of the wind. The authorial voice tells us that the Trojans shuddered when they saw the snake lying in their midst, "a portent of aegis-bearing Zeus" (12.208–9).[39] But the statement that the snake was a portent (*teras*) is syntactically so positioned as to leave it ambiguous whether this is the opinion of the authorial voice or the focalized view of the shuddering Trojans.

Polydamas, neither a seer nor priest himself, pleads with Hector to retreat. He believes that the omen means the Trojans will end up retreating in disarray, with many fatalities. He reinforces this view by claiming that this would be the interpretation of a prophet "who had clear knowledge of omens in his mind and whom the people believe" (12.228–29). But

Hector has been told, by Iris, the will of Zeus: he must remain on the battlefield until Agamemnon leaves it (11.199–209, 11.284–91). He is disinclined to believe Polydamas's interpretation of the incident of the eagle and the snake. It could just be, after all, a coincidence. He tells Polydamas that he would rather trust in the message he has personally received from Zeus than in what he thinks is a random zoological event. He continues (12.237–42):

> But you tell us to obey long-winged birds
> that I neither pay attention to nor concern myself with,
> whether they move on the right towards the dawn and sun,
> or on the left towards the murky darkness.
> Let us obey the advice of great Zeus,
> who is king of all mortals and immortals.

Hector's strategy is then endorsed: Zeus sends a blast of wind that raises a dust-cloud to confuse the Achaeans, and the Trojans "trust in his portents and in their might" (12.252–57).

Of course, both omens are fulfilled. One refers to the eventual retreat of the Trojans several books later; the other refers to the Trojans' temporary ascendancy in the imminent battle of the ships. Hector is faced with conflicting messages. Gods sometimes are and sometimes are not behind visible phenomena in the natural world.

Yet, in practice, the experts in interpreting divine signs in the *Iliad* are always proved correct. Achilles rightly points out that the plague-ridden Achaeans need to ask a seer or priest or reader of dreams, which come from Zeus, why Apollo is angry (1.62–64). Calchas has been given the gift of prophecy by Apollo, so he knows what is and will be as well as what has been (1.70). Odysseus remembers how, long ago, at Aulis, Calchas had interpreted the omen of the blood-red snake that Zeus sent from beneath the altar to the plane-tree when the Achaeans were gathering; it devoured a mother sparrow and her eight chicks before being turned into stone (2.303–32). On the Trojan side, the two sons of a priest of Hephaestus face Diomedes, and Hephaestus saves one of them by enfolding him in darkness to spare his priest a double bereavement (5.9–24). Other experts include Eurydamas, "reader of dreams" (5.149), whose two sons are killed by Diomedes. The Corinthian Euchenor was the son of a prosperous seer named Polyidus, who had predicted that if the young man went to Troy he would be killed there (13.633–70). Meriones kills Laogonos, the son of

Onetor, who was priest of Zeus on Mount Ida and honored locally "as if he were a god" (16.603–5).

Unrecognized Divine Interventions in the Physical World

There is no reason to consult a seer or priest if you are completely unaware that the gods are intervening in your life. A warrior whom you think you know may actually be a doppelgänger, like the phantom (*eidōlon*) of Aeneas that Apollo provides to divert Diomedes (5.449). A god can help your weapon find its mark, as Athena helps Diomedes's spear penetrate Ares's stomach, apparently without Diomedes knowing it (5.840–59).

A warrior in the *Iliad* ought also to question whether an apparent mortal who engages with him directly might not be a god in disguise. Zeus sends Agamemnon a deceptive dream that takes the form of Nestor (2.6–21). Athena disguises herself as Antenor's son Laodocus to approach Pandarus (4.86–88). Apollo races Achilles in the guise of Agenor (21.600–605). Aeneas wonders but cannot know whether Diomedes in full attack on the battlefield is not some god, angry with the Trojans on account of omitted or mishandled sacrifices (5.177–78); Pandarus comes nearer to the truth in believing the man to be Diomedes, protected by an immortal who stands beside him, his shoulders "wrapped in cloud," turning missiles aside (5.183–91).

Homer's heroes were of course unaware that later, after the destruction of Troy, Poseidon and Apollo would obliterate the Achaean wall by harnessing and uniting the power of all the rivers of Ida—Rhesus, Heptaporus, Caresus, Rhodius, Granicus, Aesepus, and the two river gods later to be depicted in dialogue, Scamander, and Simois; Zeus assisted these rivers in the destruction by sending continuous rainfall (12.17–32).[40] But humans are also often unaware when a god is manipulating a natural phenomenon in the moment. A pair of vultures on an oak tree might be Athena and Apollo secretly spectating (7.58–59); only the seer Helenus knows who they are because he hears their voices (7.53). Achilles is apparently not aware of Athena when she sprinkles nectar and ambrosia on him to prevent him from feeling hungry while fasting (19.352–54). The failure of a corpse to putrefy in the normal manner should be (but is not) attributed to the gods by Hecuba (24.757–58). Zeus is able to make Olympus quake by nodding in assent or taking his seat on his throne (1.530; 8.443), and so is Hera (8.199). Zeus is repeatedly said to send the rain that causes rivers to rise (5.91; 11.493; 12.25; 16.386); the scholia interpret the adjective *diïpetes*

("Zeus-sent" or "god-sent") as referring to precipitation sent by Zeus.[41] Suddenly waking or going to sleep may be a sign that a god wants to attract or divert one's attention (10.518–19, 24.445). A touch of Hermes's wand can instantly induce sleep or cause wakefulness (24.343–44).

Mist, cloud, and darkness frequently disguise interference by deities, especially on the battlefield. Poseidon and the other pro-Achaean gods convene before they do battle at the wall that the Trojans and Athena had once built for Heracles to protect him from the sea monster (which other sources say was sent by Poseidon, to whom Hesione was to be sacrificed); as they convene, they drape an unbreakable cloud around their shoulders (20.150).[42]

Aphrodite hides Paris in a thick mist to remove him from danger (3.380–81). Apollo hides Aeneas in a dark cloud (3.343–44) to do the same. Ares shrouds the battle in night to help the Trojans (5.506–8); he himself can take on the appearance, in Diomedes's eyes, of a black darkness produced by clouds when a blustering wind arises after a hot spell of weather (5.865). Zeus hides his horses in a thick mist near his sanctuary on Mount Ida (8.50). Athena can remove the cloud of mist that blinds warriors in battle (15.668–69). A sudden dust storm may be caused by Zeus, unbeknownst to mortals, by arousing a sudden wind from Ida (12.253–54). Apollo uses thick mist to befuddle Patroclus so that Hector can kill him (16.788–89). But Zeus sheds nocturnal darkness over the fights for Sarpedon's corpse to protect it (*nux*; 16.567–68). He subsequently spreads a thick dark mist (*aër*) over the corpse of Patroclus and later disperses it after Ajax prays to him to remove it (17.269, 368, 376, 644, 645, 649).

Achilles is apparently unaware that Poseidon has shed mist over *his* eyes in order to rescue Aeneas from the battlefield after Achilles's spear has penetrated his shield; Poseidon speaks directly to Aeneas, but Achilles needs to *infer* that Aeneas has been saved by the favor of the immortal gods, without knowing exactly which one (20.318–50). Agenor is unaware that Apollo is rousing him, because the god enfolds himself in deep mist surrounding an oak tree (21.549). Presumably when Agenor is forcibly snatched away by Apollo and engulfed in mist, he has some apprehension that a god is involved, although this is not made explicit (21.597). Apollo can hide Hector in a thick mist so Achilles can only stab at air (20.443–47), and can draw clouds (*nephos*) over Hector once he has expired to stop his body putrefying (23.188).

The post-antique mind has difficulty with these phenomena, which at different moments are explicitly said to be both godsent and naturally occurring. Anachronistic euhemerism, rationalization, and allegorization are always tempting. The twelfth-century Byzantine poet and grammarian John Tzetzes saw profuse allegory—rhetorical, natural, and mathematical— everywhere in the *Iliad*. "Natural" allegory, for him, satisfactorily explains the poem in terms of its climatological, etiological, and environmental features. He gives the example of Apollo as plague-bringer because the sun, the natural allegory of the god, is the actual source of the disease.[43] The Iliadic gods, for Tzetzes, are sometimes psychological characteristics like love or wisdom, sometimes monarchs or sages, sometimes planets, but sometimes, especially in the *Theomachia* of book 20, allegories of elements.[44]

Mark Edwards proposes that the idea of a sudden obscurity on the battlefield "must have arisen from the clouds of dust stirred up during an actual battle."[45] As Jonathan Fenno points out, Edwards's comment is foreshadowed by the long traditions of rationalization in, for example, Eustathius, who allegorically interpreted both the darkness of night that Zeus spreads over Sarpedon, and the mist poured over Patroclus, as battle-dust, *konis*.[46] Fenno is surely correct in suggesting that "the natural phenomenon of mist, like the psychological impulses of Homeric heroes, is subject to double causation or over-determination: physical nature (like human nature) and divinity can produce the same phenomenon simultaneously, interchangeably or independently."[47] Such mist is simultaneously "a natural meteorological phenomenon associated with water" and "a typical Homeric instrument of divine intervention manifesting Zeus' sympathy for Patroclus."[48] Drawing on Philippe Descola's *Beyond Nature and Culture*, which emphasizes the sheer diversity of ways in which humans and non-human entities have been configured worldwide across time and cultures, Olaf Almqvist has persuasively charted the fluidity between myth and philosophy in the archaic age of Greece.[49] The poets who produced the *Iliad* preceded by only a couple of centuries the pre-Socratic thinkers who initiated the formal divorce between natural phenomena and anthropomorphic gods.[50]

Elemental and Meteorological Analogies

Elements and weather are a pronounced aspect of the *Iliad*'s rich imagery: the very strife of war can "be ablaze."[51] When a god or mortal is *compared* with such a phenomenon, the effect can suggest that they effectively took on its physical appearance, as when Thetis comes out of the sea in answer to Achilles "like a mist" (1.360), or Athena darts from the peaks of Olympus like a gleaming meteoric star emitting sparks as it flies, a sign sent by Zeus for seamen or soldiers (4.73–78). Iris flies to speak with Poseidon like snow or hail driven down from the clouds by the blast of the West Wind, born of the bright sky (15.170–71). Ares bellows to the Trojans from the apex of the city, "like a murky whirlwind."[52]

Human actions are compared with weather phenomena as well. In a complex simile, Agamemnon groans repeatedly like Zeus setting off lightning as a prelude to either a storm of rain, hail, or snow, or to war (10.5–8). Diomedes sleeps outside his hut with his arms and armor, and his comrades sleep with him, their heads cushioned on their shields and their spears stuck upright into the ground; the bronze weapons shine afar like Zeus's lightning (10.150–53). Hector intermittently appears and disappears from the Achaeans' view amid the front rank of the Trojans, with his shield shining like a deadly star repeatedly reemerging from clouds, flashing like the lightning of Zeus (11.61–66). Hector's opponent Ajax attacks in battle like a river in flood arriving as a torrent on the plain from the mountain in winter, driven on by Zeus's rain, bringing with it oaks and pines, and casting driftwood into the sea (11.492–95).

But humans can also exhibit behavior that, on numerous occasions in the *Iliad*, prompts a simile comparing them to a natural phenomenon that occurs spontaneously *without* any apparent intervention of a god, or to a human confronting a random natural phenomenon. The bronze arms of soldiers on the move gleam like a limitless forest set alight by consuming fire on the peaks of a mountain; they march as if all the Earth were swept with fire; the words fall from the chest of the great orator Odysseus like snowflakes on a winter's day.[53] When the two sides collide in battle, their shouting and struggles resemble winter torrents that gush down from the mountains to a place where two valleys meet and merge their mighty floods in a deep ravine, making a thunderous din that a shepherd far away in the mountains can hear (4.452–56). Diomedes concedes ground like a man traversing a great plain who halts helplessly at a swiftly flowing river streaming out to the sea, and leaps backwards when he sees it heaving with foam (5.597–600). Agamemnon and Patroclus both weep like a spring of dark water that pours its murky stream down a sheer cliff-face (9.14–15; 16.3–4).

Agamemnon resembles the joint energy of both fire and wind when he fells Trojans like devastating flames that fall on a virgin forest and are exacerbated by a wind that whirls them in every direction, so that the thickets are flattened by the forceful onrush of fire (11.155–57). Hector, son of Priam, peer of Ares, the bane of mortals, falls on the conflict like a blustering tempest that lashes the violet-colored sea into motion (11.297–98; see also 12.40); Nestor recalls coming upon the enemy "like a black tempest" in a long-ago battle between the Pylians and the Epeians (11.747).[54]

Both Achaeans and Trojans fight the ships by casting missiles that fall thick and fast like snowflakes to the ground, hurled down by a strong wind that drives the shadowy clouds along (12.156–57). When Hector smashes the gates apart and breaches the wall, he leaps into the Achaeans' garrison "like sudden night," his two eyes blazing with fire (12.462–66). The Trojans follow him in one great mass, like flame or a tempest-blast (13.39). Poseidon, disguised as Calchas, rouses the two Aiantes against Hector (13.53–54), a berserker who leads his men on like blazing fire (13.53–54). The authorial voice compares Hector to a hard boulder detached from the surrounding rockface by the force of a river in full flood swollen by rain in winter; it leaps high, making the forests resound as it bounces down to the plain (13.137–41). Idomeneus arms himself, and goes on his way (13.242–44),

> Like the lightning which the son of Cronus
> seizes in his hand and flourishes from shining Olympus,
> revealing a sign to mortals, and the forks flash brilliantly.

When the Trojans see him, his strength is like a flame (13.330).[55] But Hector, too, can brandish his spear like a flame (20.423).

The combat between the two sides resembles the great cloud of dust raised when whistling winds meet a road thickly blanketed with dust (13.334–35). They fight "like blazing fire" (13.673), especially godlike and flamelike Hector (13.688); Hector, booming "like a snowy mountain" (13.754), rallies the Trojans around wise Polydamas.[56] The successive ranks of Trojans in their glorious assault at the climax of book 13, urged on by Zeus, are like the dreadful winds that blast the Earth when Zeus thunders, creating high-arched foaming waves that follow them across the surface of the loud-resounding sea (13.795–801). Even internal thought processes can be likened to natural phenomena. When Nestor deliberates whether to join the battle or go to join Agamemnon, before he makes up his mind his thoughts are divided like a sea that is beginning to stir as the winds rise, apparently spontaneously, but its waves do not fall in one single direction, until Zeus sends a specific gale (14.16–21).

As the battle at the ships culminates, the meteorological and elemental imagery reaches a climax. The Achaean wall stays firm, like a huge steep

sea-cliff that resists both shrill winds and swelling waves, but Hector, shining like fire, leaps on them like a wave whipped up by the winds falling on
a ship, engulfing it with foam, as the wind roars against the sail and the
oarsmen are petrified (15.618–27). When Zeus's heart is set on pleasing
Thetis by granting Hector temporary glory in the battle of the ships, he
rouses the warrior who is already eager for the fray (15.605–9):

> So Hector was raging like Ares the spear-brandisher, or destructive
>   fire
> raging in the thickets of a deep mountain wood,
> and he was foaming at the mouth, and his eyes
> glittered beneath his bristling eyebrows, and his helmet
> shook terribly around his temples as he fought.

Patroclus wonders whether Achilles was born not of Peleus and Thetis but
of the gray sea and steep cliffs, so unbending is his heart (16.33–35). When
the Danaans win a little breathing space during the battle at the ships, it is
like when Zeus shifts a cloud to reveal not only the mountain tops, headlands, and valleys, but even the "infinite air" (*aspetos aithēr*) that reaches to
heaven (16.297–300); the claustrophobic perspective of humans trapped in
a restricted arena, fighting for their lives, is suddenly replaced with a cosmic perspective of a landscape and skyscape stretching off into infinity.

During the battle over Patroclus's corpse, the Trojans shout as loud as
a river, where, at its mouth, the stream collides with sea and the headlands
on the coast resound with the roar (17.263–65); but at this point Zeus
sheds darkness over the battlefield where fighting raged around the corpse
(17.259–70). This murk is so intense that it was as if neither sun nor moon
existed anymore (17.366–67). The conflict becomes as fierce as a fire that
suddenly takes hold of a city, making its houses collapse under the might
of the wind that drives the glaring flames along (17.736–39). The two
Aiantes hold the rear defenses (17.747–51),

> like a wooded ridge
> that happens to lie at angles to a plain,
> and contains even the troublesome streams of mighty rivers,
> and instantaneously forces all their currents to wander over the plain,
> and the flow cannot force a way to break through.

Achilles, glimpsed from the wall by Priam, is like the gleaming Dog Star
that rises at harvest time (22.26–32). Hector trembles when he sees Achil-

les, from whose body the bronze shines like the light of a burning flame or of the rising sun (22.134–35). And when Achilles poises his spear ready for the death blow, the spear itself emits a gleam like the evening star in the darkness of night (22.317–18).

Wai Chee Dimock is correct in writing that the "storm surge" of violence unleashed on the battlefield unfolds, via the similes, "in the domain of physics," with the result that, in the realm of the senses, the war comes to resemble "a natural disaster."[57] Yet it is striking how little we hear about the actual elements and weather conditions at Troy. The overall sensory impression the poem undoubtedly conveys of intense elemental and meteorological activity, as Brooke Holmes points out in one of the few publications to have wrestled seriously with the ecology of the *Iliad*, is almost all through the medium of analogy.[58] Just occasionally we can sense the physical world apparently acting in synergy with human activity, in a form of "pathetic fallacy." Alone in an open space on the "wave-beaten" shore "of the loud-resounding sea," Achilles mourns Patroclus, "groaning deeply" (23.59–60), the sound of his lamentation echoed in the noises of the sea. When the fighting at the ships is at its most intense, we hear that the sea surged up to the huts and ships of the Achaeans and the two sides clashed, making a huge din (14.392–93).[59] But the actual surging of the sea almost immediately prompts a series of *comparisons* of the warriors' clamor with natural phenomena. The men shouted louder than sea waves bellow on the shore, driven by the dreaded blast of the North Wind, Boreas (14.397); louder than a forest fire roars; louder than shrieking wind (*anemos*) in oak trees (14.399).

The conjunction of the named North Wind and the unnamed wind in the oaks brings us to one more category of intermittently personified natural phenomena. An unnamed favorable wind (*ouros*) can be given by an equivalently unnamed god to weary oarsmen (7.4–6). Shrill winds (*anemoi*, 14.17) and a tempest are presaged by a heaving sea. The Trojans flood over the Achaean wall like a great breaker out at sea, when the force of the wind (*anemou*) drives it on, smashing down onto the sides of a ship (15.381–83). Hector leaps on the Achaeans like a storm wind (*isos aellēi*, 11.297). But underlined here is the difficulty of deciding whether winds are perceived anthropomorphically, because the poet develops this comparison (14.305–9):

As when the West Wind (*Zephyros*) propels the clouds
of the brightening South Wind (*Notos*), striking it with a strong
     tempest,
and many swollen waves roll on, and the spray

is scattered above by the blast of the wandering wind (*anemos*),
so many heads of the enemy contingent did Hector lay low.

Perceptible winds, of course, are themselves without substance, and this was apprehended by Homer's audiences, who understood the word *anemōlios* (windy) metaphorically, to designate empty words, boasters, or a futile action (4.355, 20.123, 21.474). Winds are envisaged as operating alongside, but independently of, an anthropomorphic god, when the dust falling on the Achaeans is likened to the chaff that useful blasts of wind (*anemōn*) pile up at winnowing time, "when yellow-haired Demeter separates the grain from the chaff" (5.499–501).

There is little attempt to visualize the appearance of even the named winds; the ancient audiences seem to have been most interested in the different effects created by winds coming from different directions; one simile details how a precipice or promontory is wave-beaten on all sides precisely because it is afflicted by winds from every direction (2.394–97). Missiles fly like winter snowflakes willed by Zeus when he *lulls* the winds, so that the snow can fall continuously to coat the entire landscape—mountains, headlands, plains, fields, harbors, and shores; the only part it does not cover is the sea itself (12.279–86). Arms and armor pour out of the ships as fast and thick as snowflakes sent by Zeus beneath the chilly blast of the North Wind (19.357–58). But most of the winds in the *Iliad* seem appropriate to a story based at the western end of a huge land mass (Asia) close to the Aegean Sea, to the west of its western coastline.

The Achaeans stir like waves raised in the Icarian Sea (west of Asia Minor) by the East Wind (Euros) or South Wind (Notos) as they drive on the clouds of Zeus in the mountainous interior (2.145–46), or by the West Wind (Zephyros) crushing the ears of corn in a field (2.147). The Achaean battalions surge like the cresting sea driven by the West Wind to crash in repeated waves onto the land (4.422–26). A "cloud" of infantry resembles a pitch-black cloud driven across the sea by the West Wind, presaging a whirlwind on land that forces a goatherd and his flock into a cave (4.275–79), an image that offers a peculiar "tension between distant vision and human vulnerability."[60] When the ranks of seated soldiers on either side move to take seats in the plain, they are like the sea that blackens as it is rippled by the newly risen West Wind (7.63–64). The Achaeans are plunged into grief like a dark cresting wave, which spews out seaweed from the fish-infested sea upon sudden blasts of the North Wind and West Wind that blow from Thrace (Boreas and Zephyros, 9.4–8).

But at other times it is implied that the named winds are physically embodied. When Sarpedon is about to expire, the "breath of Boreas" (*pnoiē Boreao*) revives him by "breathing" on him (5.697–98). Boreas and the other winds sometimes sleep, allowing mists to linger on the mountain-tops (5.552–56). The East and South Winds are imagined in combat with one another as they compete in shaking the branches of trees (16.765–69). They are envisaged as running as well as breathing or blowing: Achilles's immortal horse Xanthos says that he and Balios can run as swiftly as the West Wind, which men say is the fastest (19.415–16). Boreas assumed the disguise of a dark-maned stallion to impregnate Erichthonius's mares (20.221–25). Hera rouses Zephyros and Notos (21.331–41) to hasten Hephaestus's fire across the plain when Achilles is fighting the River; she had once been in league with Boreas, long ago, when she wanted Heracles to be driven across the sea to Cos (15.26–28).[61]

Finally, the named winds are embodied sufficiently to enjoy eating a banquet. When Achilles prays to Boreas and Zephyros to come and help him kindle Patroclus's fire, Iris hears him. She flies to Zephyros's house, which seems to be located over Thrace (23.230), where all the winds are feasting. Iris halts at the stone threshold; as soon as they see her, each wind calls her to come to him, perhaps implying libidinous intent (23.200–204); this detail presumably gave rise to the later tradition that she eventually married Zephyros.[62] The winds obey her instruction, go to Troy, and fall upon the pyre, beating its flames all night long; at dawn they finally return home (23.214–30).

Analogies with Wild Fauna

There are perhaps faint residual traces in some epithets of an earlier, totemistic religious phase: ox-eyed Hera (8.470), owl-eyed Athena (1.206), wolf-born Apollo (4.101, 119). A subsequent discussion of farming will address the many similes in the *Iliad* involving domesticated livestock, sometimes threatened by wild animals: more than half the poem's similes, 125 out of the total 226, feature animals, whether tame or wild.[63] Dogs fall into a special category of their own. But wild fauna unassociated with domesticated animals—mammals, birds, aquatic creatures, reptiles, and insects—share the world inhabited by the heroes of the *Iliad*, too.

The mountain range of Ida is not only well provided with springs but is also "mother of wild beasts" (8.47).[64] Pandarus's treacherous bow was made by a craftsman from the long horns of a wild ibex that Pandarus had

ambushed (4.105–11). The Boeotian regions of Thisbe and Messene are
"abundant in doves" (2.502); a dove is the target in the archery contest
(23.854–56). There are fears and threats expressed that heroes' corpses
will be devoured by vultures (4.237; 22.42); Diomedes berates Paris with
the taunt that any man who comes against him will be surrounded "by
more birds of prey than women" (11.395).[65] Menelaus warns Euphorbus
against boasting: his estimation of himself is higher than that of a leopard,
lion, or furious and mighty boar (17.19–24).

But, like weather and the elements, wild fauna feature far more heavily in the similes than in the central narrative, and the similes involving
them steadily become more elaborate and complex over the course of the
poem.[66] The Achaeans assemble like swarms of bees over flowers in spring
(2.87–89) and pour from their ships and huts like different species of large
aquatic birds—wild geese, cranes, and long-necked swans (2.459–63). The
comparison illuminates the troops' pride in their strength, their movements in all directions, and the noise of their shouting. Their sheer numbers are then compared with those of huge swarms of flies that buzz around
dairies in spring (2.468). The Trojans' clamor sounds like the calling of
cranes escaping wintry storms and measureless rain (3.2–5). Athena wards
off an arrow from Menelaus like a mother dashing a fly off a sleeping child's
face (4.130–31). Agamemnon accuses the Achaeans of standing around
dazed like exhausted fawns (4.243–45). Achilles says he used to labor in
battle to secure gains for others rather than himself, like a mother bird
who brings food for her chicks but does not look after her own needs
(9.323–24). Ajax retreats slowly, dazed, "like a wild beast" (11.546). Meriones springs forth like a vulture (13.531). Thetis, Athena, and Apollo each
spring down from Olympus "like a falcon" (18.616, 19.350, 15.237–38).
Menelaus looks for Antilochus, scanning the entire Achaean force on the
battlefield, like an eagle, "which they say has the sharpest sight of all winged
creatures under heaven," as it spots a hare from on high and swoops down
to seize it (17.474–81). In the boxing match at the funeral games, Euryalus
springs up and collapses again like a fish that leaps and is then concealed
by a wave under the ripple caused by Boreas (23.692–93).

Fighting itself prompts many wild animal comparisons. As a battle is
joined, both sides leap on one another or rage "like wolves" (4.471, 11.72);
Diomedes, Odysseus, Agamemnon, and Achilles are also all compared in
short simile phrases with lions (10.297, 11.239, 24.572). Lion similes may
be short or long, but there are more lion similes in the poem—fifty—than
any other category.[67] When the Trojans become ascendant, they charge
against the ships like flesh-devouring lions (15.92–93).

Paris is compared with a snake in the mountain that is terrified when a man stumbles upon it (3.33–35); Hector awaits his last battle like a snake that has eaten toxic plants of his own accord, or has been poisoned by human guile, and wrathfully awaits the man, staring terribly as he coils himself in his nest (22.93–95).[68] But hostile relations between humans are also likened—in much longer, more developed "nature red in tooth and claw" similes—with predation of one kind of wild creature on another.[69] The dominant scholarly view these days emphasizes the savagery of wild animals rather than their reciprocal vulnerability to humans, holding that the *Iliad* uses its similes to show how "civilised values" dissolve among humans when they become heated enough in their lust for bloodshed on the battlefield.[70] My own view is that emphasizing violence between wild animals obscures how much violence is directed toward them by human beings, savagery that rarely receives comment from academics.

The most grisly description of such violence is reserved for the climactic moment when the Myrmidons are finally unleashed by Achilles. They charge from their tents, fully armed (16.156–63),

> like carnivorous wolves, whose courage is limitless,
> when they have cut down a great horned stag in the mountains
> and devour him, and the jaws of all of them are scarlet with blood;
> they go in a pack to lap with their narrow tongues
> the black water from the surface of a murky spring,
> belching out bloody gore. The hearts in their breasts
> do not tremble, and their stomachs are distended.

The "meaning" of this simile depends, however, on whether the focus is on the wolves or the stag. Humans at other times surround stags and tear them to pieces, too.

Several extended similes compare a warrior in combat with a lion attacking another wild animal: Menelaus, seeing Paris approach, is as glad as a hungry lion that alights upon the carcass of a stag or wild goat (3.23–26); the Trojans are unable to save Isus and Antiphos, two sons of Priam, from Agamemnon, any more than a doe can save her fawns from a lion when he has taken them from their den (11.116–19); Hector and Patroclus struggle over the corpse of Hector's charioteer Cebriones, like two hungry lions competing for a dead deer (16.756–58).[71] But Hector is soon to be victorious, and kills Patroclus like a lion killing a mountain boar when they compete over a meager trickle of springwater (16.823–26). The relationship between a divinity and a mortal can occasionally be likened to a lion attacking prey: Hera warns Artemis not to attempt combat with her, since Zeus "made her a lion against women, allowing her to kill any one of them she wanted." Artemis would do better to stick to killing beasts in the mountain than to take on beings more powerful than she (21.483–86).

Finally, interaction between humans in conflict is repeatedly characterized by a simile comparing it with the pursuit of smaller birds by predatory ones. With Zeus's assistance, Hector charges toward the ships like a tawny eagle swooping down onto wild geese, cranes, or swans by a riverbank (15.690–92). Patroclus, infuriated by the loss of the Myrmidon Epeigeus, charges against opponents like a falcon putting doves and starlings to flight (16.581–83). Automedon in his chariot swoops on the enemy like a vulture on a flock of geese (17.461–69). The Achaeans flee like squawking starlings or jackdaws when they see a falcon approaching (17.755–57). Achilles pursues Hector like a shrilly crying falcon chasing a trembling dove (22.139–42). Just once is such a simile used of one deity fleeing before another, when Artemis flees in tears from Hera's beating, like a dove before a falcon (21.493–95). The level of her humiliation is perhaps suggested by this unusual use of the predatory bird comparison to characterize an altercation between immortals.

(Meta)physics

In the *Iliad*, humans are aware of their own acute vulnerability to environmental and meteorological events. They know that the relationship between themselves, the anthropomorphic gods, and natural phenomena is intricate and often difficult to decipher. In the case of gods and natural phenomena, the boundary between corporeality and being constituted by a sentient force field or immaterial entity is permeable. There is an apprehension of a numinous divinity lurking behind some grammatically feminine but not explicitly personified natural phenomena such as the sky, sea, and land, since they are called "divine" (*dios*). Some natural phenomena seem on occasion to be anthropomorphically embodied and sometimes not: Dawn, Helios, Earth, winds, and rivers.

It is possible for a god to be indistinguishable from an element he often harnesses; Hephaestus both uses and can represent fire. A rainbow may mean that a god is sending a message via Iris, but it might just be a random phenomenon or indicate the presence of Athena. Any event within humans' terrestrial physical environment that is discernible to them may or may not be a sign of divine activity. Humans acknowledge that even the expert specialists in interpreting, for example, bird omens, may be misguided. A sign unanimously agreed to emanate from a single god, such as thunder and lightning that belongs to and is sent by Zeus, can mean different things to different witnesses; success or failure, approval or criticism, prediction of an outcome that is imminent or will be long delayed.

The boundaries between divine, human, and even animal and elemental spheres are strikingly permeable. Mortal and immortals mingle, as do mortal and immortal horses and rivers. Gods mate with mortals and mortals may be half divine, a quarter divine, one-eighth divine, or more distantly descended from a god. A substantial group of warriors are children, grandchildren, or great-grandchildren of river gods or nymphs. A hero does not need divine ancestors, however, to receive the patronage of a god, and there is a large range of vocabulary to suggest that humans are dear to gods, like gods, equivalent to gods, "sprung from Zeus," or the recipients of specific gifts from gods, whether inborn excellence in a particular sphere, status, or a physical object.

Humans may know that a god is manipulating a natural phenomenon, but they are frequently unaware of it when gods disguise themselves as vultures, put them to sleep or wake them up, or send mist, cloud, and darkness. The difficulty of knowing whether a god is behind a physical or environmental event is emphasized in the extraordinary variety of similes comparing figures and actions, in the main narrative, to phenomena in nature. We hear little about the actual elements and weather conditions in the main narrative, and instances of pathetic fallacy are few. Untamed fauna of all kinds share the world of the heroes, but like weather and the elements, preponderate in similes.

Humans in similes may be loggers, farmers, or metalworkers, and nature is also vulnerable to them. As König puts it, "the similes contribute perhaps more than any other aspect of the poem to the *Iliad*'s place as a major landmark in the Western environmental tradition."[72] The world of the similes also portrays chaotic complicated interactions, involving metal tools and weapons, between humans, trees, and domesticated livestock as well as creatures of the wild. This dialectical complexity is compactly evoked in a great simile when Aeneas slaughters the Achaean brothers Crethon and Orsilochus, descendants of the River Alpheus (5.554–60):

> They were like two lions raised on the peaks of a mountain
> by their mother, in the thickets of a deep wood,
> and they grab cattle and plump sheep,

plundering the farmsteads of men, until they, too,
are killed by sharp bronze held in human hands.
That was how these two were defeated beneath Aeneas' hands,
and they fell like tall fir-trees.

The confusing and phantasmagoric ferocity of the *Iliad*'s interactions between humans, nature, and divinities can be read as testimony, however aesthetically mediated, to an embryonic intuition of the scale of the damage that human activities might inflict on the cosmic ecosystem in its entirety.

CHAPTER FIVE

Loggers

It started with the catalogue of ships:
Whole forests felled for keels, masts, spars, oars, hulls
Made black and waterproof with tar and pitch.

—A. E. STALLINGS

MOUNT IDA LOOMS OVER the skylines, ecology, and rituals of the *Iliad*. The mountain, a long subalpine massif consisting of limestone and green and black schist blocks, lies in what is now northwestern Turkey, twenty miles southeast of what are believed to be the ruins of Troy at Hissarlik. Poseidon, who knows about woods, since he has a sacred grove on Onchestus, northwest of Thebes (2.506), describes Mount Ida as "wooded" and "with many ridges" (21.439).[1] The Trojan Agenor, under pressure on the battlefield, intuitively considers going to hide in the thickets of the mountain (21.559). Many streams and waterfalls trickle down its southern slope, a phenomenon that gave rise to another of Ida's Homeric epithets, "with many fountains" (for example, 14.283).[2] On the northern side, running westward to the sea, is the River Scamander that provided ancient Troy with its fresh water. The mountain is also called in the same line "mother of wild beasts." Deer, wild boar, and jackal still roam Ida today, but the wolves, lynx, brown bears, and big cats have long since been wiped out by overhunting.

These epithets for Ida appear at a crucial moment in the *Iliad*, when Hera needs to seduce her husband so as to divert his attention from the battlefield. She goes with the god Sleep to the heights of Ida; when they reach the highest part of the forest, Sleep climbs the very highest fir tree and disguises himself as a bird. The tree would have been an evergreen coniferous Turkish Fir (*Abies nordmanniana*, the subspecies labeled *equitrojani* because it was, sensibly enough, believed by botanists to have provided the wood for the Trojan Horse). This species, which is endemic to this single small location on Ida, has been degraded by consistent human logging for at least four thousand years, and is now officially endangered.[3] It can reach a height of sixty-one meters, which explains why Homer said that this tallest tree "reached through the mists up to heaven" (14.288).[4] Hera proceeds up to the highest summit at Gargarus (now Karataş), which rises a soaring 5,820 feet (1,770 meters) above sea level.

There Zeus sees her, and they make love after Zeus conceals them in a dewy, glistening golden cloud. The bed provided for them is a floral *locus amoenus* in a description of natural beauty rare in the poem (14.347–50):

> The divine earth put forth fresh new grass beneath them,
> and dewy lotus and crocus and hyacinth,
> thick and soft, which lifted them from the ground.

Even in the Bronze Age, the summit of Ida, without such poetic intervention, was probably exposed and bare. Farther down the mountain slopes there is still an abundance of botanical species—succulent grasses and colorful flowering plants—thought to have been left there after the last Ice Age.[5] But these plant species, like the trees and animals, are dwindling rapidly as a result of human activity and global warming; some are under threat of extinction.[6]

The gods are drawn to Mount Ida. It is one of Zeus's favorite personal domains, and Gargarus contains his fragrant altar (8.47–48).[7] Troy can flourish because of the rains that fall and the rivers that rise on its ridges (12.19–22). According to Aeneas, Dardanus, the first ancestral king of the Trojans, built his city, Dardania, "on the ridges of many-fountained Ida" (20.216–18). The next two kings, Erichthonius and Tros, remained there. It was Tros's son Ilus who established the new city of Ilios on the plain. The man who moved down from the mountain is a minor but significant part of the *Iliad*'s backstory, and the tomb of this "leader of the people in old times" outside the city walls remains an important landmark.[8]

When Achilles selects a particular tree stump as the turning post in the chariot race, Nestor assumes it will be of oak or pine, types of timber that do not rot in the rain (23.327–28). There are two other important trees providing landmarks outside Troy. One is a wild fig tree (6.433, 11.167, 22.144), not far from the wall, and the other is a tall wild Valonia oak in the plain. Apollo and Athena meet at this oak, disguising themselves as vultures to sit on it and watch the fighting (7.60).[9] It is not far from the Scaean gates (11.170), and it is probably meant to be understood as the same as the beautiful oak that Sarpedon's comrades make him sit beneath when he is wounded (5.693), and the oak on which Apollo leans, wrapped in mist, to encourage Agenor (21.549).[10]

A tamarisk tree, growing somewhere between the Achaean camp and that of King Rhesus of Thrace, is also significant in the narrative. On its branches Odysseus leaves the spoils taken from Dolon, as a marker to enable Diomedes and him to find their way back again through the dark night (10.465–67). Perhaps it is the same tamarisk that Adrastus's chariot becomes entangled in, an accident that leads to his death at Menelaus's hands (6.39).

The importance of timber to the economy assumed in the *Iliad* is reflected in some of the epithets applied or alluding to other places. For Michael Williams, these signify that Homer's audience already felt that particularly dense wooded areas were precious, rare, and remarkable.[11] Perhaps they are evidence that there was already wide concern about conserving timber supplies, but, although that concern would surely be expected, it is never made explicit in the *Iliad*. Andromache's mother, before being captured by Achilles, had been queen in "wooded Plakos" (5.425). Poseidon's favored island of Samothrace is likewise "wooded" (13.12). The Carian mountain Phires, some way down the coast from Ida, is "dense with foliage" (2.868). Odysseus's Cephallenians hold Ithaca and Neriton "of the waving forests" (2.631–32), a lovely epithet that also applies to Pelion in Thessaly (2.757).[12] Achilles's own enormous spear was made of ash from a tree on Mount Pelion (22.133); the forest-dwelling centaur Cheiron had given it to Achilles's father Peleus "for the slaughter of warriors" (16.141–44). Mount Pelion's supplies of ash were depleted long ago. Until recently it had survived as one of the last densely forested places in Greece, but in the summer of 2007, sadly, a fire destroyed a large proportion of its remaining trees. The extent to which climate change has contributed to the increasing number of wildfires in Greece has yet to be determined conclusively by scientists, but local people living in the area of Mount Pelion are convinced that it is a significant factor.

Long before the Bronze Age, Stone Age hunter-gatherers had already set fire to woods to open up meadows so that they would attract prey.[13] In the Mesopotamian *Epic of Gilgamesh*, King Gilgamesh's fellow warrior and beloved friend, a wild man of the hills named Enkidu, moves to the city after helping Gilgamesh kill Humbaba, the giant who guarded the forest; they cut down the cedars. The Israelites used massive amounts of cedar logs to build the Temple of Jerusalem.[14] But it was the ancient Greeks and Romans who transformed the timbered landscapes of the Mediterranean and Black Sea worlds. Russell Meiggs's canonical study *Trees and Timber in the Ancient Mediterranean World* demonstrates at length and in erudite detail just how important wood was to almost every dimension of "classical" civilization. Navies, pasturage, farming, mining, carpentry, and smelting "voraciously devoured forest."[15] Timber met 90 percent of fuel needs.

Large-scale deforestation accelerated between about 600 BCE and 500 CE, but the irrevocable damage had begun in the Bronze Age. Timber shortfalls probably contributed to the collapse of power in Minoan Crete, and the Mycenaeans depicted in the *Iliad* made things worse by clearing land for sheep grazing and arable farming. Pollen samples show that pine forests grew around Pylos on the Peloponnesian coast, the homeland of the Iliadic hero called Nestor, in the sixteenth century. But nearby Messenia had at least four hundred smiths turning out many tons of metal. The pottery industry expanded enormously and needed vast amounts of fuel.[16] As the economy expanded, so did the population in the late Bronze Age, especially the thirteenth century. It is generally held that at that time there were as many as two million inhabitants of the territory encompassed by Greece today.

Pollen studies and soil profiles record the toll on the forests of the Peloponnese caused by unprecedented economic and population growth. Palynology shows that in Greek mountain areas the vegetation patterns were already changing in the Bronze Age (second millennium). The coastal pine forest around Pylos was destroyed and turned into pasturage; woodcutters had to move to the center of the peninsula by the thirteenth century BCE. So did potters, who relocated to mountain slopes.[17] The Cycladic island of Melos, where the Mycenaean housing at Phylakopi had been constructed out of timber, was entirely deforested.[18]

Such deforestation must have been connected with the natural disasters that became much more common during the late Bronze Age. The newly treeless slopes above the Plain of Argos sent earth and water raging downhill in the rainy season. Mud-filled torrents and drought afflicted Tiryns and Navarino (Pylos) harbor, eroding the soil and leaving it short of nitrogen to this day.[19] Drought also became more common. By the third century CE, there was no extensive forest anywhere in the plains and low hills around the shoreline of the Mediterranean. Classical civilization wrecked the environment. Theophrastus knew that the microclimates had changed. He was, moreover, interested in the relationship between species of plants and the environments in which they could flourish. Not only did his use of the word *oikeios* in reference to what is conformable with the nature of a particular species lie behind our modern word "ecology"; he is also regarded as the founding father of Forest Science.[20]

"Infinite" Wood for Funeral Pyres

It is the trees not of mainland Greece but of Ida that are consumed in gargantuan quantities in the action of the *Iliad*. At the poem's final emotional climax, after forty days of conflict, brutality, and emotional agony, Priam asks Achilles for permission for his Trojans to leave the city, inside which they are pent up, to gather wood from far away on the mountain (24.662–63). On his return, he orders the Trojans to collect wood for the dead Hector's pyre. It is to be a great pyre, fitting for the best of the Trojan warriors. The men of the city go out to the mountains and "for nine days they collected immeasurable amounts of wood" (24.795).[21]

The Greek word "immeasurable" here, *aspetos*, literally means "too great to be spoken of," "ineffable" or "infinite." Homer uses it elsewhere to describe cosmic elements—the sky, the stream of Ocean, the waters of the sea (8.558, 18.403, 23.127), while the early philosopher Empedocles used it to describe the infinity of time.[22] Homer's choice of epithet reveals a secret about the worldview of the eighth century BCE, when his epics reached their final form. Timber was seen an infinite resource, and as such could be consumed in immense quantities to make a statement at a funeral. But timber is not infinite. We know that now. The UN Environment Programme's 2020 report, *The State of the World's Forests: Forests, Biodiversity, and People*, makes for a bitter and terrifying read.[23] The world loses at least 37 million acres of forests every year, exacerbating the greenhouse effect.[24] And although there is no evidence that the ecological devastation has eased over the past thirty to forty years, the endangerment of the Amazon rain forest attracts less attention than it did in the 1980s and 1990s.[25]

Timber is also used earlier in the poem, when Andromache issues her futile order to her women to heat water in a cauldron so that Hector can bathe on his return (22.442–45). Nor is Hector's funeral the only one in the poem requiring abusively aggressive logging. There are other obsequies, beginning with the densely packed pyres that burned for nine long days in the Achaean camp as Apollo relentlessly let loose his plague-bearing arrows at the mules, dogs, and especially, the people (1.50).[26] For the funeral of Patroclus, says Achilles, the Achaeans need enough wood for an "inexhaustible (*akamaton*) fire," to ensure the flesh of the corpse is fully burnt (23.152), and thus release the soul to Hades. Agamemnon orders the Achaeans to climb the spurs of "Mount Ida with its many fountains" and chop down a vast number of "high-crested" oak trees to make a pyre (23.110–28):

> But Lord Agamemnon
> urged on mules and men from all the huts
> to fetch wood. And the noble Meriones,
> squire of manly Idomeneus, supervised them.
> They set out, holding axes in their hands to cut wood with,
> and well-plaited ropes. The mules went on ahead of them.
> They moved frequently up, down, aside, and aslant.
> But when they reached the shoulders of Ida with its many fountains,
> straightaway they applied themselves to felling high-crested oaks
> with long-edged bronze. The trees kept on falling
> with a mighty crash. Then the Achaeans split them and bound them
> behind the mules, which tore up the earth with their hooves
> as they made eagerly for the plain through the thick undergrowth.
> All the woodcutters were carrying logs as ordered
> by Meriones, squire of manly Idomeneus.
> Then man after man hurled them down on the shore, where Achilles
> indicated should be the great barrow for him and Patroclus.
> And when they had thrown down the immeasurable
> wood on all sides, they sat down there in a throng.

The first timber to be burned during Patroclus's obsequies will heat the cauldron of water that will wash the dead hero's corpse (18.343–48):

> With these words, godlike Achilles told his comrades
> to put a great tripod over the fire, so that they could swiftly
> wash the clots of blood from Patroclus' body.
> And they placed on the blazing fire a cauldron for the bathwater,
> and they poured the water in, and took planks of wood and set fire to
>   them.

Timber for Patroclus's pyre. Drawing by Becky Brewis of a
nineteenth-century engraving that depicts the funeral of Patroclus.
(Courtesy Becky Brewis)

The fire flickered around the belly of the cauldron and the water
    heated up.

Agamemnon has dismissed most of the Achaeans, leaving only those clos-
est to Patroclus to perform the cremation (23.163–65):

The mourners remained there and piled up the wood,
making a pyre a hundred foot square;
with grieving hearts they placed the corpse on the top of the pyre.

Watchfires, Roasting, and Burnt Sacrifices

In book 7 there are two mass funerals for the dead of both sides, which
involve the collection of even more wood (7.417–32):

Then the Trojans prepared themselves with great swiftness for both
    tasks,
some retrieving the corpses and others collecting timber.
And on the other side the Argives hastened from their well-benched
    ships,
some retrieving the corpses and others collecting timber.
When Helios began to strike the fields,
as he rose to heaven from the gentle flood
of deep-flowing Ocean, the two sides encountered one another.

It was a hard task then to identify each man individually.
But they washed the clots of gore off with water,
shedding hot tears as they lifted them onto the wagons.
Priam would not allow weeping, so his men heaped the corpses
on the pyre in silence, their hearts breaking.
And when they had burned them in the fire they went to sacred
    Ilium.
On the other side, the well-greaved Achaeans did the same.
They heaped the corpses on the pyre, their hearts breaking.
And when they had burned them in the fire, they went to their
    curving ships.

But fires are lit for other combat-related reasons than funerals. There are watchfires and foundry fires that burn profligate amounts of timber. In book 8, Hector orders the Trojans to gather an abundance of wood to make hundreds of watchfires (8.507–9). Moreover, every house in Troy is to build a great fire (8.520–21) to ward off ambushers. The fires, we hear, burned in multitudes (8.555–63):

It was like when in heaven the stars twinkle with great clarity
around the shining moon, when the air is windless,
and all the mountain-peaks are revealed, and the steep headlands
and the valleys; and the infinite air breaks open all the way to heaven
so all the stars are visible and the shepherd is glad at heart.
In such numbers gleamed the fires that the Trojans kindled
in front of Ilium, between the ships and the streams of Xanthos.
A thousand fires burned on the plain,[27]
and by each one sat fifty men in the glow of the blazing fire.

Rather than *chili'* in line 8.562, "a thousand," Zenodotus, the great Homeric scholar and librarian at Alexandria, read *muri'*, which could mean "numberless" or more specifically "ten thousand" here, making a total of half a million warriors sitting around them![28] But it is just as important that the earthbound view of the shepherd seems combined with a more cosmic scale of perception, foreshadowing the alarming fusion of human and divine perspectives portrayed on Achilles's shield.[29]

In book 9, the Achaeans built a fire large enough for seven hundred youths, plus seven sentinels, each armed with a long spear, to cook their individual suppers over its flames (9.85–88). The heroes of the *Iliad* behave, therefore, as though wood is an infinite resource. But this is nothing new even in the world of the *Iliad*. The elderly Phoenix remembers how as a young man, when his people feared he would escape from a difficult domestic situation, they kept watch over him for nine whole nights. They kept the watchfires in front of the "well-fenced wooden" courts lit for all those nine consecutive nights (9.470–72).

Wood is destroyed to provide fuel for cooking and burnt sacrifice as well. Eating raw meat is cast rhetorically as an act of primitive barbarism in the *Iliad*, and timber is needed in particularly large quantities to roast large animals. The River Scamander seethes like the water in a cauldron when dry faggots are placed under it and set alight to melt lard (21.362–64).[30] Every time mention is made in the *Iliad* of an animal being cooked, sacrificed, or presented to the gods as a burnt offering, it assumes the destruction of trees to produce firewood. Pasturing livestock itself threatens forests and woodland. But feeding human carnivores requires a continuous supply of timber, too.

For context, an Argentinian restauranteur who specializes in roasting whole cows, in a solemn ritual, needs 250 logs (almost a ton) to feed the fires for the twenty-three hours it takes to roast a single whole cow weighing a thousand pounds.[31] Slightly less wood would be required if the dead animal is jointed and some of the meat placed on spits, the procedure described in the *Iliad* (2.421–29):

> When they had prayed and sprinkled the barley-grains.
> first they pulled the animals' heads back and cut their throats and
>   flayed them,
> then they cut out the thigh-meat and wrapped it
> in a double layer of fat, and topped it with raw flesh.
> These meats they burned on planks which had been stripped of
>   leaves,
> but the innards they skewered on spits and held over flaming
>   Hephaestus.
> But when the thigh-meat was grilled through and they had tasted the
>   innards,
> they jointed the rest and put it on skewers,
> and roasted it carefully, and pulled it all off the skewers.

The poet requires his audience repeatedly to imagine vast bonfires being lit to roast sacrificial animals in honor of the gods. For each animal slaugh-

Animal sacrifice to a goddess. Drawing by Becky Brewis
of a red-figured pelike in the Kunsthistorisches Museum,
Antikensammlung, Vienna, inv. 1144. (Courtesy Becky Brewis)

tered, his audience knew that several trees needed to be cut down, made
into suitably shaped planks, and denuded of foliage.

Chryses reminds Apollo that he has a good record of sacrificing bulls
and goats to him (1.40–41). Achilles wonders whether Apollo has found
fault with a vow or hecatomb and may require lambs and goats to be sacrificed to him (1.65–67). Calchas prescribes that, if they are to be released
from the plague, the Achaeans must return Chryseis to her father without
ransom and lead a sacred hecatomb to the island of Chryse (1.98–100). In
the event, they offer hecatombs of both bulls and goats (1.315–16), so that
the savory smell went up to heaven, "eddying amid the smoke" (1.317).[32]

Odysseus remembers the day in Aulis when an omen appeared while the Achaeans were offering hecatombs (2.305–6). In the episode preceding the catalogue of ships, it is implied that all the Achaeans made individual sacrifices to the gods over fires in their huts (2.399–401). Although it is impossible that there was an animal killed for every single soldier, as the Greek suggests, the cavalier attitude toward numerical accuracy implies a conviction that resources were infinitely replenishable. Agamemnon, aware that his problems after the altercation with Achilles are caused by Zeus (2.375–76), sacrifices a five-year-old bull to him and prays that he may take Troy that very day (2.402–10). Zeus accepts the sacrifice, "but did not yet grant him fulfilment," instead making the agony "unceasing" (*amegarton*, 2.418–20). In Athens, we are told, the youths make regular sacrifices of bulls to the mythical king Erechtheus to win his favor; he was born of the "life-giving land," but fostered by Athena (2.546–49).[33]

Agamemnon sacrifices the lambs at the ritual oath-taking (2.292–94), but the perjurious Pandarus promises Apollo a hecatomb of first-born lambs (4.119–21). Routine eating of meat may be confined to the elite, monarchical class (12.319), but the phrase "equal feast" after a sacrifice implies that the cooked meats produced at such rituals were shared by everyone in the community (4.48). It is repeatedly stated that the Trojans, especially Priam and Hector, have always been assiduous in their performance of animal sacrifice and libations to the gods (4.44–49, 22.171–73, 24.34–35), both on Mount Ida and on the citadel of Troy. When the Corinthian hero Bellerophon arrived in Lycia, the king entertained him for nine days, slaughtering nine oxen (6.174–75). The Trojan priestess Theano promises Athena that, if she helps them, the Trojans will sacrifice twelve heifers that have never felt the goad (6.308–9).

After a hard day's fighting and sunset, both sides hold banquets. The Achaeans "slaughtered oxen throughout the huts and took supper," while the Trojans feasted in the city, as did their allies: the Achaeans procure shiploads of wine from Lemnos, for which they have exchanged bronze, iron, hides, cattle, or slaves (7.466–77). Agamemnon claims that, on the voyage to Troy, he never sailed past an altar of Zeus without stopping to offer burnt beef to the god (8.238–41). One reason for the Trojans' watchfires in the plain is to roast hecatombs of sacrificial oxen and sheep on fires fed "with abundant wood," sending forth a sweet savory smell to the gods, even though the gods, hostile to Troy, refuse to partake (8.545–48).

Patroclus cooks a feast for the Achaean ambassadors to Achilles's tent, consisting of the backs of a sheep and a goat and the chine of an enormous, fatty hog (9.206–8), but he also, when Achilles instructs him, performs a sacrifice to the gods and casts burnt offerings on the fire (9.218–20). Phoenix remembers how his people had tried to persuade him to stay after his father took against him for sleeping with his concubine; they slaughtered many sheep and cattle and pigs in a feast apparently lasting for nine nights running (9.466–70). He recalls feeding meaty morsels to the infant Achilles (9.488–89), and that Artemis had once been wrathful with the people of Aetolia because she was snubbed by them when the other gods were receiving hecatombs (9.534–36). Hector used to feed Astyanax marrow and lamb fat (22.501–2). Diomedes promises Athena that, in return for her assistance, he will sacrifice a yearling heifer that has never been yoked to her (10.292–94). Nestor recalls the Pylian army sacrificing, on a single occasion, a bull to Achelous, a bull to Poseidon, and a heifer to Athena (11.727–29); when he once visited Peleus, the Thessalian king was burning for Zeus the fat thighs of a bull that the young Patroclus and Achilles were butchering (11.771–74). Agamemnon tells Talthybius to make a boar ready, to sacrifice to Zeus and the Sun (19.196–97).

Both sides conduct sacrifices to win the favor of Zeus (10.46), but the Achaeans and Trojans are not the only humans to cultivate the gods with burnt offerings: Iris wants to go to the land of the Ethiopians at the streams of Ocean, because she wants her share of the hecatombs they are sacrificing to the immortals (23.205–7). At Patroclus's funeral "many bulls," "many sheep and bleating goats, and many pigs with white tusks, rich in fat" were sacrificed (23.30–32; see also 23.166–69). Meriones is victorious in the archery competition because, unlike his adversary Teucer, he has promised to sacrifice a hecatomb of firstling lambs to Apollo (23.863–65, 23.872–73).

Nestor tells his son Antilochus that a woodsman who can use cunning intelligence as well as physical force will excel at his craft, as will a helmsman endowed with *mētis* when his ship is buffeted by stormy winds, or a charioteer who employs his intelligence (23.315–18). But the most potent celebration of a skilled craft worker in the *Iliad* is reserved for smelting operations, in the portrait of the smith god Hephaestus when Thetis finds him in his workshop, "sweating as he went to and fro at the bellows" (18.32), and "moving the bellows aware from the fire" in order to talk to her (18.412). As he sets to work, he returns to his bellows (18.468–70),

and turned them towards the fire and ordered them to work.
The bellows, all twenty of them, were blowing on the melting-vats, emitting forceful blasts of every kind.

In reality, a standard blacksmith working alone, like Hephaestus, would use only one large set of bellows. That the god has twenty apparently automatic sets, which do not require operators, conveys the enormous size of the fire alight in his forge. In the real world, although it is not here specified, the fire would have required vast amounts of timber and charcoal, made from timber.

Wooden Arms, Fortifications, Ships

There is a great deal of other timber in the actual action of the *Iliad*, most of it produced simply for the purposes of war and liable to swift and wasteful destruction, like the innumerable ashen spears (5.655, 5.666, 8.552), oaken axles of chariots that speed across the plain (5.838), or the olive-wood shaft of Peisander's axe, which is long and well-polished (13.611–13). The mass funerals are immediately followed by the building of the Greeks' wall, complete with gates and wooden pikes planted densely in the peripheral trench (7.437–41):

> They built a wall to defend their ship and themselves,
> and inserted into it gates that fastened tightly,
> to provide a carriageway through it for horse-drawn chariots.
> Outside it they dug a deep trench,
> broad and large, and they planted stakes in it firmly.

The entire wall is topped with huge, sharp, thickly set stakes that frighten Hector's horses (12.55–57, 12.63–65), but the entire edifice is vulnerable and easily destroyed with crowbars (12.256–57). On board their ships the Greeks have additional pikes of timber, tipped with bronze, in case the fighting reaches the decks (15.388–90); the one Ajax wields is particularly spectacular, "a long pike for sea-fighting, jointed with rings, twenty-two cubits in length" (15.677–78).[34] Since the Greek cubit was equivalent to about eighteen inches, this makes Ajax's weapon for sea fighting no less than an unbelievable thirty-three feet, or ten meters, long.

There were at the outset far more than a thousand Greek ships, each equipped with dozens of benches and oars made of polished fir (7.4–6), and all held in place on the beach at Troy by wooden prop mechanisms (*hermata*, 2.155). The vast number is emphasized by the inclusion of the information that in a previous assault on Troy, Heracles had taken the city with "only" six ships (5.641), and their size is evoked by the epithet "huge of hull" (8.222). The amount of wood required to make even one ship was considerable. Oars and masts were made of single trees. By classical times, from which data is more abundant, it can be approximated that a trireme would require at least fifty cubic meters of worked wood. This translates into two hundred trees, each four to five meters high, plus, for oars and masts, about fifty-five trees more than twelve meters high. A rough estimate suggests that one acre of pine left to grow for four decades, with the trees tightly packed, and well looked after, might be able to produce enough timber to equip two to three fifty-oared vessels (penteconters).[35] The point is not that a fleet of this size was constructed in real life during the thirteenth century BCE. The point is that Homer required his audience members to imagine such an enormous exercise in timber use, revealing their assumption that the world timber supply had no limits.

Other epithets for the ships reveal the craft work that went into making them: well-benched, curving, hollow, as well as simply *black*. Meriones kills Phereclus, son of Tecton, Harmon's son, the master carpenter loved by Athena, goddess of handicraft, who had also built for Paris/Alexander "the shapely ships, source of evil, that became the bane of all the Trojans" (5.59–63). There are five shipbuilding similes that, as Janko has shown, portray wood for ships at serial stages in their manufacture—as trees cut down, as logs dragged down a mountainside, and as planks being crafted into a functioning ship.[36] During the prolonged battle at the ships, the adversaries were so evenly matched that the effect resembled the even marking made by a carpenter's measuring tool as he straightens a plank of a ship (15.410–12). Yet, in one of the *Iliad*'s rare acknowledgments that objects made from natural elements such as wood are vulnerable to destruction, Agamemnon uniquely observes that in the nine years that the Achaeans have been at Troy, the ships' beams have begun to rot and the ships' cables to unravel (2.134–35).[37]

Their huts are not tents but elaborate wooden edifices: the Myrmidons built for Achilles a "lofty hut," hewing fir beams, roofed with thatch gathered from the meadows, surrounded by a great court with thick-set pales and a door barred by a single, enormous fir log that takes three ordinary men to pull open or shut; only Achilles can do it alone (24.448–57). Wood is used in all kinds of architecture; one simile compares wrestlers' grips on their opponents with the gale-rafters of a house joined together by a renowned craftsman (23.712). Priam's high, vaulted treasure chamber is made of fragrant cedar wood (24.191–92; see also 24.317–18).

Men as Trees, Trees as Men, and
Forces That Destroy Them Both

Warriors are likened to trees that survive buffeting by weather.[38] The two Lapiths defending the Achaean fleet stand firm, "like high-crested oaks in the mountains, that always withstand the wind and rain every single day, firm fixed with huge unbroken roots" (12.132–34). The two Ajaxes restrain the enemy like a wooded ridge that holds back even mighty rivers (17.746–48). But more often the warriors themselves, almost all on the Trojan side, are compared at the moment they are killed to trees being felled. When Ajax kills the Trojan youth Simoeisios, his victim "fell to the ground in the dust like a poplar tree that has grown in the meadow of a great marsh, smooth-stemmed, but with branches growing from the top. Some wheelwright felled it with gleaming iron to bend a curving wheel for a beautiful chariot, and it lies drying on a river's banks" (4.482–87). This emotive simile certainly enhances the pathos of the young man's death; Schein has argued that his loss is thereby rendered symbolic of the cost of life, especially Trojan life, in the war as a whole, because the death is framed in botanical language. Simoeisios's father is named Anthemion, or "Flowery," and the youth is described as "blooming" (4.474). The simile therefore expresses the idea of the waste of tender, animate human potential.[39] But surely it also emphasizes the loss to the meadow of the "smooth-stemmed" poplar tree, felled with gleaming iron.

Aeneas kills the Achaean twins, Crethon and Orsilochus, and when they fall they are compared in a brief simile to lofty fir trees (5.560). Teucer kills Imbrius, who "fell like an ash-tree on the peak of a mountain that can be seen from a great distance around, when it is cut down by the bronze and lowers its tender foliage to the ground. That is how he fell, and around him clanged his armour" (13.178–81). When Idomeneus kills Asius, "he fell as an oak falls, or a white poplar or a tall pine, which carpenters cut down in the mountains with newly sharpened axes to make a ship. That was how he lay stretched in front of his horses and chariot, groaning and clutching at the bloody dust" (13.389–93). Elizabeth Minchin sees a contrast between the positive, productive labor of the lumberjacks and the wasteful killing of humans, but that is to overlook the pathos inherent in tree-felling.[40] Poseidon enables Idomeneus to stab Alcathous by immobilizing him; Alcathous stood fixed "like a pillar or a high and leafy tree" (14.537).[41] In this simile, Alcathous is vulnerable to felling, like a tree, but Idomeneus is implicitly presented as a man who cuts it down. Even Hector,

when he briefly collapses, is compared to an oak tree uprooted by a lightning bolt, emitting the stench of sulphur (14.414–16).

In the long central sequence of fighting that ultimately leads up to the
deaths of Sarpedon and Patroclus, the similes from the timber industry,
and other similes evoking natural disaster destroying forests, accumulate
apace. It is as if the rows of men are somehow fused with the rows of trees
that needed to be destroyed to make international military confrontation
possible.

These images create an aural crescendo of insistently hammering axes
and crashing tree trunks until the death of Sarpedon. After Patroclus's death,
Menelaus slaughters Euphorbus, "as a man rears a robust olive sapling in
an isolated place, where ample water bubbles up, a fine sapling which flourishes, and the gusts of all the winds make it vibrate, and it teems with white
blossom; but suddenly a wind comes upon it with a forceful hurricane and
tears it from its furrow, and stretches it out on the ground" (17.53–58).

All the heroes compared with felled trees at the moment of their deaths,
except for the twins Crethon and Orsilochus, are on the Trojan side.[42] The
similes cumulatively evoke a picture of ruthless Achaean loggers hacking
down a forest in a land far from home, reminding us that the *Iliad*, like the
Odyssey, was in development during the supposedly "dark age" of the eleventh to eighth centuries BCE.[43] Greek (mostly Ionian) settlers were arriving from mainland Greece on the Asiatic seaboard, where they founded,
among other cities, Phocaea, Priene, Miletus, Ephesus, Colophon, and
Clazomenae. As mentioned earlier, this movement eastward is by convention labeled the period of Greek "migrations" rather than "colonization,"
to distinguish it from the larger-scale, politically more sophisticated expansion across the Mediterranean and Black Sea that followed in the later
eighth century. But not all the people the "settlers" encountered will have
welcomed them, nor will they have always been keen to share their natural
resources with newly arrived Euboeans, Phocians, Thebans, Athenians, and
Peloponnesians. These seagoing Greeks will have required and expropriated a constant supply of Anatolian timber.

The sense of the affinity between felling men and felling trees is pro-
jected rather differently in a circumlocution for a time of day, "afternoon"
(11.84–90):

As long as the dawn lasted and it was early in the sacred day,
the missiles cast by both sides found their target, and the people kept
    falling.

> But at the time of day when a man who fells oaks prepares his meal
> in the mountain glades, when his arms become weary
> of cutting down tall trees, and he feels exhausted in spirit,
> and he longs in his heart for delicious food, at that time of the day
> the Danaans, through their bravery, broke through the battalions.

Timber production penetrates to other realms of metaphor. Paris compares his unsympathetic brother's unyielding heart to "an axe that is driven through a beam by the hand of man who skilfully shapes a ship's timber, and it adds to the force of his blow" (3.61–62). When Hector and Paris arrive to support the Trojan militia, the sight is as delightful to the Trojans as the fair wind that a god gives to oarsmen when they are exhausted from striking the sea "with polished oars of fir wood" (7.4–6). The gleam of the bronze in battle is like that of a "limitless forest" ablaze in the mountains (2.455–58). The din of battle, the clanging of weapons, sounds like the din, heard from afar, when lumberjacks hack down oaks in the glades of a mountain (16.633–37). The men crashing into one another become like trees themselves (16.765–69): "As the East Wind and the South strive with one another in shaking a deep wood in the glades of a mountain—a wood of beech and ash and smooth-barked cornel, and these dash their long boughs against each other with a wondrous din, and there is a crashing of broken branches."

When the Achaeans try to retrieve the corpse of Patroclus, they are likened to mules "that drag down from the mountain a great log, or a piece of timber for a ship" (17.742–46). Patroclus's inanimate body is no longer a living tree, but a chunk of matter. The comparison of men to trees reaches its most tragic articulation, twice within a few hundred lines, when Thetis refers to Achilles's impending death. She tells first the Nereids, and then Hephaestus, that after she had borne a son without equal (18.56–60; 18.437–41),

> he grew up fast like a sapling.
> Then, when I had raised him like a tree in a rich orchard plot,
> I sent him out in the beaked ships to Ilium,
> to wage war on the Trojans. But I will never again
> welcome his return to the house of Peleus.

Thetis's flourishing young tree will all too soon be felled.

Warriors are also likened to natural forces (some of which as we now know can be exacerbated by human activity) that spell ruin for woodland. Agamemnon, on the rampage, fells the heads of fleeing Trojans (11.154–57), "as when obliterating fire falls on a forest that has never been logged, and the whirling wind carries it in every direction, and the corpses fall, torn up by the roots by the force of the onrushing fire."[44] Ajax, raging on the battlefield, is likened to a river in flood that sweeps away trees (11.492–95), "as when a flooding river comes down into a plain, a winter torrent from the mountains that Zeus's rain drives on, and it carries many dry oak trees and many pine trees, hurling abundant debris into the sea." Polypoetes and Leonteus fight like wild boars being pursued in the mountains—boars that crush the trees around them, knocking them down at the roots, as they charge from side to side (12.145–49).

The combined war cries of Achaean and Trojan forces, noisier than the waves of the sea under the onslaught of the North Wind, were also as loud as the roar made by a blazing forest fire whose flames leap as they incinerate a forest in the mountain glades, and as strident as a wind roaring through the high crests of oak trees (14.396–99). Hector rages like a consuming fire that seethes in the thickets of a deep wood (15.605–10). In the battle over Patroclus's corpse, the clamor arose from both sides "as the din of men who cut down oaks in the glens of a mountain, and a noise arises that can be heard from afar. That was what the clanging was like as it arose from the broad-wayed earth, a clanging of bronze and well-wrought shields" (16.633–37).

On rare occasions, the *Iliad* seems to imply that trees have a special relationship with humans, for example the elm trees planted round Eëtion's barrow by mountain nymphs (6.419–20). Hector says to himself that he can no longer dally with Achilles as "from oak-tree or rock" (22.126–27).[45] Although it is equally possible that this proverb, having arisen from Minoan and Mycenaean divination practices, is about types of language rather than human genesis, some scholars have seen here traces of a tradition that members of the human race are descended from trees; in *Works and Days*, Hesiod says that the men of the Bronze Age were sprung from ash trees.[46] But there is little sense in the *Iliad* that trees are animate or even organic entities, let alone an apprehension of speaking dryads. In Chaucer's courtly *Knight's Tale*, the woods and trees of the landscape serve primarily "as a backdrop to human affairs"; they are dispensable.[47] Similarly, in the *Iliad*, trees may be splendid, but their purpose is to be exploited by human beings.

Humans are thus cut down like trees, sound like trees, grow up like trees, but also threaten those trees, destroy forests, and run like wildfire wrecking an entire forest. Achilles rages like a dreadfully blazing fire running through the deep vales of a parched mountainside, as it is driven on, whirling; the black earth ran with blood (20.490–94). As we shall see later, in the description of his shield, Achilles's fight with Scamander, and the River's fight with Hephaestus's fire, the implication of man in nature, as simultaneously its adversary, its wrecker, and its victim, becomes crystalized as the very symbol of the Anthropocene, of *Homo sapiens*'s troubled, irreconcilable relationship with the natural world.

CHAPTER SIX

Farmers

ONE REASON THAT ANCIENT forests were cleared was to make land available for arable farming and the grazing of livestock. As Lucretius puts it, earlier humans "used to force the woods daily ever further up the mountains, conceding the lower lands to cultivation."[1] The type of tree that flourished in a particular soil was regarded as an indicator of the crops that could flourish there instead. Trees were cut down with the saws and axes that appear in some Homeric similes, and in the account of Odysseus felling twenty tall alders and firs near the shoreline to make his boat to escape from Calypso's island (*Odyssey* 5.234–44). Loggers used axes up to six feet long, saws with dense rows of teeth, and guiding ropes to aid felling. Large trees were girdled and left to die off before being chopped down.[2] Their roots were customarily removed and often then burned, and the fertilizer produced by their ashes was valued highly.[3]

When Aphrodite is injured by Diomedes, she emits ichor, not blood, because the gods do not eat bread nor drink wine, so they are bloodless and called immortals (5.339–42). One Homeric definition of mortals is that they cultivate plants to eat, "the fruit of the field" (6.142), especially cereal crops.[4] Idomeneus says that Ajax, son of Telamon, will yield to no man "who is mortal and eats the grain of Demeter, and can be torn apart by bronze or great stones" (13.322–23). Viticulture, reaping, baking, and vulnerability to death by metal thus define the mortal human. But, in a more heightened ritual phrase, "eating the grain of Demeter" seems to be a synecdoche for eating anything at all; Lycaon, threatened with death by

Achilles, and supplicating him, reminds him of his particularly sacred obligation to spare his life, because it was at his table that Lycaon had partaken of "the grain of Demeter" (21.76).

In just one dazzling simile, we are asked to visualize golden-haired Demeter herself separating grain from chaff on a windy day at winnowing time (5.499–501). When the fillies whom the North Wind fathered on Erichthonius's mares (20.226–27) galloped over the grain-giving land, they would skim the topmost ears of ripe grain without damaging them (20.226–27). Phthia, Larisa in the Troad, Lycia, Paeonia, and several other locations are "deep-soiled" (1.155, 17.172, 2.851, 17.350).[5] The Perrhaebi come from the land that they work around lovely Titaressus (2.751). The sanctuary of Demeter is in fertile land at flowery Pyrasus (2.695). Although the *Iliad* overlooks the actual hard labor and muscle power that grinding grain demanded (acknowledged in the *Odyssey* twice—6.103–4 and 20.105–8—where it is done by females marshaled in large teams on Phaeacia and Ithaca), the stone that Ajax crashes down on Hector was the size of a millstone (7.270).

These relatively infrequent evocations of arable farming bring glimpses of an intersection between the *Iliad* and another of the few great surviving archaic Greek hexameter poems, Hesiod's *Works and Days*. Demeter at winnowing time is an image reminiscent of Hesiod's advice to set slaves to winnow "Demeter's holy grain" on a smooth threshing-floor in an airy place.[6] Hardly has Hesiod's instruction of Perses begun, when the authorial voice stresses the imperative of growing and accumulating a sufficient quantity of "Demeter's grain" to last a year.[7] Perses is warned that only agricultural labor will save him from famine,

> so that hunger (*limos*) detests you
> and revered Demeter of the fine crown loves you,
> you must fill your barn with food;
> hunger is altogether a suitable companion for the idle man.[8]

Perses needs to "do work upon work on work," by stripping off his clothing to plow, sow, and reap every single year, if he and his family are not to end up begging for food.[9] He is given detailed advice on how to identify suitable trees for felling to make a cart: he needs no fewer than a hundred sections of timber.[10] Perses also needs suitable oxen to pull his plow and a phlegmatic middle-aged journeyman to drive it.[11] From his farmstead in a "miserable hamlet" in the foothills of Mount Helicon, he might have had ample reason to challenge the assumption in the *Iliad* that it was always easy for a lone peasant to find and fell "infinite" timber.[12]

The meteorological world conjured in the similes of the *Iliad* is, similarly, the actual world of the Hesiodic farmer; the poets of war and of peace shared the same stock of weather formulae. Perses is to beware of the frosts,

> which come cruelly
> over the earth when the North Wind blows;
> he blows from across horse-breeding Thrace
> onto the wide sea, and stirs it up; earth and forest howl.
> He falls upon numerous high-leafed oaks and bushy pines,
> lowering them to the all-nurturing earth in the mountain ravines.
> Then all the infinite forest roars.[13]

We have a clue to the way in which Hesiod might have heard some of the *Iliad*'s weather similes when he uses almost identical language to Homer's, but in a recommendation to take suitable sartorial precautions against rain:

> When the frosty season arrives,
> sew together the skins of firstling goats with ox-hide tape,
> to put over your back and provide protection
> from the rain. Put a shaped felt cap on your head
> to keep your ears from getting wet;
> the dawn is cold under the onslaught of the North Wind.
> At dawn a mist that nurtures wheat is spread over the earth
> from starry heaven, on the fields where men that are blessed labour.
> It is drawn from the ever-flowing rivers
> and lifted high above the earth by a squall of wind,
> and sometimes towards evening it turns to rain,
> and sometimes to wind when Thracian Boreas stirs up the thick clouds.
> Complete your work and get home before he arrives.
> Never let the dark cloud from heaven soak you through,
> making your body damp and drenching your clothes.[14]

Curiously, the sensible Hesiodic notion that clothing needed to be weatherproof, hardly a preoccupation of the *Iliad*, is shared in the war poem by Achilles's mother, Thetis. She is a sea goddess, one of the fifty Nereids, daughters of the ancient sea-god Nereus, but she was briefly married to a mortal, Peleus, before abandoning him. She remained a constant advocate of the son she had conceived, and plays an important role in the *Iliad* as his supporter. The clothes she packed for her son when he went to war at windy Ilium included "cloaks to keep out the wind" (16.224).[15] The composers and performers of the Hesiodic poems were familiar with much of the poetry that constitutes the *Iliad,* and vice versa. Hesiod believes that in the age of heroes, the age preceding the age of iron he lives in, some Greeks died when they sailed to Troy for Helen's sake; he has himself only ever sailed once, but it was to Euboea from Aulis, where the Achaeans marshaled their armed forces before the Trojan expedition.[16]

The rare, sudden flashes of Hesiodic insight intruding into the supererogatory properties of the *Iliad* offer a clue as to how some of the more skeptical of Homer's earliest listeners may have responded to the idiom of quantitative hyperbole in which most of the poem is expressed. But if the anonymous *Contest of Homer and Hesiod* is anything to go by, the sheer supererogation of the world depicted in the *Iliad* and of its poetry is part of what made it so enjoyable.

The surviving manuscript text of the *Contest* stems from a fourth-century work, the *Mouseion* of Alcidamas, a pupil of Gorgias. The fundamental idea of an *agōn* between verses of Homer and verses of Hesiod is certainly of earlier origin, perhaps the sixth century.[17] Such a contest underlies the rhapsodic competition at the end of Aristophanes's *Peace* and the tragedians' confrontation in *Frogs*.[18] The important point is that, although the king presiding over the contest overrules the popular vote, insisting that Hesiod's poetry is more useful, the audience does not want to reward utility. What they like is to be amazed by Homer's "golden" verses. They continue to prefer Homer after Hesiod asks how many Achaeans went to Troy, and the poet of the *Iliad* answers, "by means of an arithmetical problem":[19]

> There were fifty fire-hearths, and in each one
> fifty spits, with fifty pieces of meat on them,
> and thrice three hundred Achaeans round one piece of meat.[20]

There is then what is probably a Byzantine comment intruding in the text here,

> This works out as an incredible quantity, for if there are 50 hearths, the spits come out at 2,500, and the meat pieces at 125,000, so the number of men would be 112,500,000.[21]

But the poet of the *Contest* knows that improbably enormous quantities do not spoil the pleasure offered by the *Iliad:* instead, they enhance its impact.

At the climax of the contest, when the audience is already calling out for victory to be awarded to Homer, King Paneides ("Know-All") asks both poets to perform what they regard as their finest passage. Hesiod recites his advice on plowing, sowing, and reaping, from *Works and Days,* while Homer quotes two passages from the battle at the ships, featuring a plethora of shields, helmets, plumes, long spears, the glint of bronze, breastplates, more helmets and even more shields. The internal audience are in no doubt whatsoever: "Once again the Greeks were struck with admiration for Homer, praising the way the verses transcended the merely fitting."[22] The poem stuns and delights the audience because it goes far beyond what is required, plausible, or "real."

Plant Cultivation

The *Iliad* depicts a social order consisting of a small elite class and a very separate, larger group of peasant farmers and laborers. Ernst Gellner identified this as typical of "Agraria," or the model of settled farming societies manifested during Bronze Ages worldwide.[23] But the mode of production in the poem is more specific than this: it is on the cusp between the Bronze and Iron Ages. On the early end of the period between 1100 and around 560 BCE, iron tools, especially the iron plowshare, became widespread among Greek agriculturalists, allowing much denser and more heavily timbered soil to be domesticated for growing grain.[24] And although the *Iliad* is primarily concerned with warfare, a fertile source of description, imagery, and comparison is the plowing, sowing, reaping, and processing of crops. When the two sides reach a stalemate and are hurling stones at each other, the missiles, likened to snowflakes falling on a windless winter's day, cover the mountains, lofty headlands, "the grassy plains, and the rich land men labour over" (12.281–83).

The humans in the *Iliad* cultivate several different types of plants for food. There are orchards for growing fruit (9.540–42). Epidauros and Pedasus in the Peloponnese are covered in vines (2.561, 9.152; see also 5.91), as is Anatolian Phrygia (2.184).[25] Arne and Histiaea are both "rich in grapes" (2.507, 2.537).[26] A great reward offered by the Aetolians to Meleager, if he succeeds in killing the ravening boar that endangers them, is that he should choose a beautiful fifty-acre tract of land in the part of the Calydonian plain where the soil is most fertile, half of it suitable for viticulture and half of it consisting of cleared but as yet unplowed land (9.577–80).

Other passages point out that land ownership is strictly related to wealth and class status—a potential source of conflict also indicated on Achilles's shield. Diomedes of Argos boasts that he descends from a noble lineage, because his father, Tydeus, had lived in a household richly stocked with the means of livelihood; he owned wheat-bearing fields, well-planted orchards, and substantial numbers of livestock (14.121–24). Sarpedon reminds his fellow Lycian king, Glaucus, that they need to be seen to lead from the front because they enjoy many privileges, including huge estates on the banks of the River Xanthos containing both orchards and wheat-producing plowland (12.313–14). But soon afterward, these two landed aristocrats find themselves in deadlock: they cannot break the line of Greeks defending the wall that protects their ships, but neither can the Greeks repel the Lycians (12.421–23):

> It was like when two men, with measuring rods in their hands,
> dispute the boundary-stones in a jointly owned field,
> wrangling in a tiny patch of land about obtaining equal shares.

Here is another sudden flash of pragmatic Hesiodic perspicacity. This situation resembled the lived experience of most of Homer's listeners far more than did the aristocrats' sumptuous estates by the riverside. The class angle is emphasized by another simile shortly afterward, where the deadlock is likened to the evenly balanced scale-pans of an indigent woman weighing the wool she must spin "to earn a meagre wage for her children" (12.433–35).

The Trojan spy Dolon, dressed in a wolfskin and a cap of ferret fur, darts over the plain and overtakes his Greek pursuers as far as "the range of mules in ploughing—for they excel oxen at dragging the jointed plough through deep virgin soil" (10.351–53). During the battle at the ships, the two Ajaxes collaborate in holding their defensive ground "like two wine-dark oxen straining with one accord at the jointed plough in virgin soil, and profuse sweat oozes up from the bases of their horns; they are kept apart only by the polished yoke as they struggle along the furrow to cut through to the end of the field" (13.703–7). The greater Ajax, the son of Telamon, is shortly after likened, in another simile, to a far less docile domesticated quadruped, when, with great difficulty, he has been forced to cede some ground (11.588–62):

> He was like an ass passing beside a field when he worsts some boys—
> a lazy ass, against whose ribs many cudgels are broken.

He goes into the deep grain and mows it down. The boys hit him
    with cudgels,
but their strength is trifling. It is with great exertion
that they drive him out, and only when he has eaten his fill of
    fodder.[27]

In cereal-crop contexts, therefore, heroes are compared with both coerced
but obedient, and coerced but disobedient, farm animals.

They are also compared with reapers. Combatants facing their enemies in line fight like "reapers, pressing down the furrow opposite one
another in a rich man's wheat or barley field, and the sheaves fall thick and
fast" (11.67–69). They can be compared also with the cereal crops themselves; the Achaean forces stir like the ears of corn in a deep cornfield that
bow down as they are blasted by the West Wind (2.147–48). Even a missile
can be likened to the produce of crops being forced from their husks: when
Helenus strikes Menelaus with an arrow, it glances off his breastplate, "as
black beans or chickpeas leap off a broad spade on a great threshing-floor,
driven by the shrill wind and the strength of the winnower" (13.588–90).
Agent and acted upon, subject and object: the *Iliad* explores every permutation.

Pasturage

Forest was destroyed relentlessly for arable farming. But it was also removed to make way for grazing livestock. Thrace is "fertile and the mother
of sheep" (11.223). Boeotian Orchomenus is "rich in flocks" (2.605), Iton
is "mother of flocks" (2.696), and Argos is a "pastureland of horses" (6.152).
Odysseus resembles a thick-fleeced ram who paces through a flock of white
ewes (3.197–98). The importance of rich pastureland is reflected in formulaic epithets including "grassy Haliartus," "grassy Hire," and "deep-soiled
Antheia" (2.503, 9.150–51, 9.293–94), as well as the homes of "many-
flocked" and "many-herded" men (9.149, 9.296).

Arable and livestock farming are linked when Zeus sends Athena
wrapped in a brightly colored cloud to encourage the Achaeans. The effect
is as when he sends a lurid rainbow as a portent of war or chilly storm; it
stops men from working the land and bothers the flocks (17.547–52). Cattle and other domesticated quadrupeds—cows, goats, sheep, pigs, horses,
and other equids—are central to the economy of the world depicted in
the *Iliad*. Agamemnon has won enormous wealth, including gold, through

winning competitions with horses (9.123–27); Achilles runs as swiftly as a champion racehorse that wins prizes for its owner (22.22–23). The prize offered to an Achaean volunteering to go and spy on the Trojan camp is one black ewe with a lamb, "an incomparable possession," from every single one of the more than a thousand ships (10.214–16); one wonders how the winner would ever be able to transport this immense flock back to Greece.

Cattle played a particularly pivotal role in the hierarchical societies of the eastern Mediterranean during the Bronze Age. They "were the engines of Bronze Age agricultural systems," plowing fields "and moving bulk commodities from fields to storage facilities and markets."[28] Cattle labor was indispensable to the complex political systems that emerged in both Anatolia and Mycenaean Greece. Cattle "became a metaphor for abundance, and elites amassed enormous herds as physical manifestations of their rank."[29] Cattle often appear as a form of currency in commodity and/or service exchange and in estimations of value. Every one of the hundred golden tassels on the aegis borne by Athena is itself worth a hundred oxen (2.449). Achilles sold Lycaon on Lemnos for the price of a hundred oxen, and Lycaon purchased his freedom back by paying three times that much (21.79–80). The picture of the vast numbers of cattle sacrificed in both Homeric epics probably bears little relationship to reality.[30] In one cultic building found at Methana in the eastern Peloponnese around 150 Mycenaean terracotta figurines have been found, almost all of which are bovines; they are thought to have been substituted for real animal sacrifice since cattle were of such enormous value.[31]

Cattle rustling is assumed to be part of the heroic way of life: Achilles argues that he was better off at home, where nobody stole anything from him, and the example of theft he gives is cattle rustling (1.154–57). He later repudiates Agamemnon's offered gifts, because he can simply acquire cattle, sheep, tripods, and horses by robbery (9.406–7). Nestor recalls a major incident in his youth, created when he raided Elis for cattle and drove away no fewer than fifty herds or flocks each of cattle, sheep, pigs, and goats as well as 150 mares, many with foals at the teat. If each herd/flock is assumed to have contained the three hundred animals specified in line 11.697, this makes over sixty thousand animals from just one region in the northwestern Peloponnese! Since three hundred sheep need fifty-four acres of grazing land, and horses and cattle require more, the Elean hero whom Nestor killed is depicted as failing to guard animals that required over sixty thousand acres—more than ninety-three square miles—

of pasturage.[32] Defending that large a territory would not have been an easy task.

Herd animals were often pastured in woodland, where cows, goats, and sheep could be fed on foliage when other fodder was in short supply: farmers even sought out for them the parts of forests where the undergrowth and foliage were richest, while pigs were set to forage on the forest floor for acorns, chestnuts, and beech nuts.[33] The trampling of the forest floor by all these domesticated herd animals degraded the soil and the ability of any plants or saplings to take root again.[34]

Theophrastus, a botanist and younger colleague of Aristotle at the Lyceum, observes that among the ways of damaging trees, "cropping by animals is also deleterious, because it scorches simultaneously with cutting and removing, making the trees suffer more."[35] Herd animals did more than merely devour the trees: somehow their destructive foraging made the forest less sustainable and trees less likely to grow again. Varro was struck by the paradox that while cattle put behind the plow helped to produce both food for humans and fodder for livestock, "grazing cattle do not produce what grows on the land but tear it off with their teeth."[36]

The herds that did the most damage fastest were always goats, which, given the choice, prefer woody bushes and low trees. In a fragment of a comedy by the fifth-century Athenian dramatist Eupolis called *Nanny-Goats*, the chorus of goats describes their diet:

> We feed off every sort of tree: fir tree, prickly oak, and strawberry tree, munching on their tender shoots, and also the foliage: the medick tree and fragrant sage, and leafy bindweed, wild olive, lentisk, ash tree, poplar, holm oak, oak tree, ivy, heather, willow, thornbush, mullein, asphodel, rockrose, deciduous oak, thyme, and savoury.[37]

Virgil describes how crowds of goats seek out food in leafy woodlands even on the inaccessible heights of mountains in Arcadia.[38]

The damage caused by pastoralism is "that it makes permanent what destruction goes before."[39] Goats are particularly damaging to woodland, because they have such an appetite for the easily uprooted young saplings that spring up in land cleared by either lumberjacks or fire. They effectively prevent the renewal of arboreal life. And herdsmen actively destroyed forest by setting it alight to replace it with grassland, as Virgil describes in a simile illustrating the wave of courage washing over an army:

It was like when a shepherd sets fires at intervals across the forest,
when the longed-for summer winds rise and blow,
and the spaces between suddenly catch alight,
and a single blazing fiery line extends over the wide plains,
and he sits there, looking down on his conquest and the triumphant
     flames.[40]

In the *Iliad*, however, no such direct connection is drawn between the activities of herdsmen and the wholesale destruction, within minutes, of an entire forest. Perhaps this is an indication of the greater consciousness of the dangers of deforestation gained between the archaic and the Augustan eras.

The *Iliad* is little interested in dairy farming, although the linguistically diverse cries of the Trojans are compared with the bleating of myriad ewes who belong to an extremely rich man and who are waiting to be milked (4.433–36), and warriors thronging around Sarpedon's lacerated corpse are on one occasion compared with flies buzzing around full milk pails in a farmyard (16.641–43). Animals were, however, farmed for products other than food. In the masculine world of the *Iliad* we hear little about the production of wool—in itself theoretically a sustainable resource—despite the great tapestries woven by Helen and Andromache in their Trojan palace chambers and Aphrodite's abortive attempt to assume the disguise of an old woman who used to card wool for Helen in Sparta (3.385–9). The rare exceptions include the simile about the indigent woman quoted earlier; another simile compares the ease with which Hector lifts a boulder to the ease with which a shepherd lifts a ram's fleece in just one hand (12.451–52); on one occasion, too, a twisted woolen sling can double as a bandage for an injured hand (13.599–600). The chest of textiles that Thetis gave Achilles to take to war contained tunics and woolen rugs (16.220–24). After Hector's death, Andromache decides to burn all Hector's lovely clothing, finely woven by women's hands (22.510–14).

Most cattle in the action of the *Iliad* are there to be sacrificed and eaten, but the leather provided by their hides is itself a valuable commodity. In one of the most powerful similes in the entire poem, the enemies fighting over Patroclus's corpse are likened to a group of people standing in a circle around a great oxhide, stretching it to its maximum extent in every direction while it is still moist with fat (17.389–94). A bull's hide can be a prize in a running race (22.159–60). The single most impressive leather object is Telamonian Ajax's great shield, like a city wall, with its seven layers of oxhide, topped with a layer of bronze. We even hear the name of its creator, Tychius, the "best of workers in hide" who had "wrought it with toil" (7.219–23). Sarpedon's great bronze shield had been made by a smith, but he had stitched many oxhides inside it with golden thread (12.294–97). Hector and Aeneas proceed shoulder to shoulder, carrying their tough, dry leather shields topped with bronze (17.491–93).

The strap that holds Paris's helmet on his head is made from the hide of a slaughtered ox (3.375), as are the string of Pandarus's bow (4.122), some small round shields, and the fluttering aprons of untanned leather attached to them (5.452–53). The rim of Hector's great shield is lined with black hide that slaps against his ankles and neck as he moves (6.117–18). In the nocturnal world of the Doloneia, Diomedes and Odysseus both wear leather helmets rather than bronze ones; while Diomedes's helmet has neither peak nor crest, but fits snugly around his skull, Odysseus's features a plaited leather lining, an inner lining of felt, and studs made out of boar's teeth (10.257–65). Oxhides can be used as bedding too; Diomedes sleeps with his body on an oxhide and his head on a woven rug (10.155–56). Patroclus's attendant spreads hides on the ground as a makeshift hospital bed for the wounded Eurypylus (11.843).

One of Ares's epithets is "warrior with a tough shield of hide" (5.289, 20.78). A poetic shorthand for the occurrence of many battlefield fatalities is "many shields of bulls' hide and many helmets fell in the dust" (12.22–23). And hides can be used directly in combat. When the Trojans succeed in creating breaches in the Achaeans' wall by taking crowbars to the parapets and supporting pillars, the Achaeans respond by using bulls' hides to seal the battlements and continuing to shoot from them (12.258–65). When Achilles continues his war on Hector even after his opponent's death, he pierces his ankles so that he can thread oxhide thongs through them with which to drag the corpse behind his chariot (22.396–99).

Varro describes the conversion of arable farmland into pasturage, in a nostalgic vision of the first city-dwellers, who kept a harmonious balance between the cultivation of flocks and that of crops, only to have the balance wrecked by the greed of their descendants.[41] Similar "nostalgic pastoral" contexts are one of several types in which shepherds appear in the *Iliad*. In these contexts, Homer recalls the happier existence in the past of a warrior, usually on the Trojan side, dying on a distant battlefield when he used to herd his livestock on the hills of his homeland.

Achilles had once taken two sons of Priam captive when they were herding flocks in the vales of Ida, and bound them with the suitably pastoral flexible bonds made of long willow shoots (11.104–6), but the youths had been ransomed and returned to Troy. Hector chides his kinsman Melanippus, who, before the enemy arrived, "used to pasture his shimmying cattle in Percote" on the southern shore of the Hellespont (15.546–48). Achilles reminds Aeneas that he once chased him down from the mountains of Ida when he was herding cattle (20.188–89). This is an allusion to an episode recorded in the *Epitome* of the fourth book of Apollodorus's *Bibliotheca*; Achilles, with some of the Achaean chiefs, laid waste the countryside, and made his way to Ida to rustle the cattle of Aeneas. But Aeneas fled. Achilles killed the cowherds and Mestor, son of Priam, and drove away the sacred kine.[42]

Further back in time, pastoral scenes appear in recollections of the conception or birth of heroes that can be romantic and erotic as well as nostalgic: Euryalus kills Trojan twins whom the nymph Abarbaraē bore to Laomedon's son Boukolion (little cowherd); the nymph had conceived them when he was shepherding his flocks and she lay with him (6.24–26). Telamonian Ajax strikes Simoeisios, "whose mother had once given birth to him as she came down from Ida to the banks of the Simois, where she had followed beside her parents to keep watch over their flocks" (4.474–76; the Simois, now called the Dümrek, is in our era imperiled by soil erosion).[43]

Throughout the *Iliad* the poetic evocation of infinite resources, and colossal consumption, usually obscures the harsh realities of physical life where large numbers of humans and animals are competing for these resources. But, as we have seen, there are a few exceptions, flashes of candor that sound less like Homer's than those of Hesiod, the grumbling peasant farmer. One fascinating passage reveals the Homeric poet's awareness that the spatial constraints imposed on both sides participating in the war will have made access to pastureland and fodder especially valuable. The perjurious Trojan aristocrat Pandarus explains the reason that he has left all the twenty-two horses that drew his eleven chariots in their own stables at his father's home, chomping away on barley and wheat (5.202–3): "I was sparing my horses, to prevent there being a shortage of fodder in a place where men are closely confined; the horses are accustomed to an adequate diet."

Similar concern for horses' welfare is expressed when we hear that Andromache used to serve Hector's horses with wine and grain before she served him (8.186–90), that Patroclus used to wash the manes of Achilles's horses in bright water and then smooth them down with olive oil (23.280–82), and that Menelaus fears that his horses will be hurt at the funeral games (23.434–37). But Pandarus's prosaic concern is altogether different. His unusual account throws a disturbing light on the vast numbers of domesticated livestock supposed to be kept alive and, as we saw earlier, sacrificed, cooked, and eaten in the *Iliad* by both sides. Hesiod insists that, after harvest, Perses should bring inside ample fodder and litter for his cattle and mules.[44] If Hesiod had listened to the *Iliad*, he would, like Pandarus, have asked where exactly all the animals were pastured and stabled. Did the Trojans and Achaeans climb higher and higher up the slopes of Ida in pursuit of food, fodder, and places to accommodate livestock, to areas made newly accessible by all the loggers' depredations?

Domesticated animals appear in numerous similes in the *Iliad*; in the simplest, the situation depicted is of an idealized rustic normality in which the humans on the battlefield are likened to animals firmly under human control. One of the famous formulaic phrases, "herdsman of the people," used of Agamemnon and of other military leaders, seems to have generated the simile at the marshaling of the Achaean troops, when "the chieftains assembled them on each side for entry into battle, as goatherds separate easily the scattered herd of goats when they join them in the pasture" (2.474–76). But immediately after this, in another simile, the relationship between greater and lesser humans is illustrated by the role of the alpha male bull in the herd rising conspicuously taller than the rest of the cattle, as Agamemnon stands out among the other warriors (2.479–81). One beautiful simile applied to both Hector and Paris asks the listener to imagine a domesticated animal breaking free from human control altogether. When Hector hears Apollo encouraging him, he runs into the conflict like a stalled horse at the manger that breaks its halter and charges over the plain to bathe in the river, reveling in its own beauty as its mane flows over its shoulders (15.263–68; for Paris, see 6.506–11).

Human-animal interaction colors two famous similes where the Trojans are likened to flocks of sheep under the management of man. Because they speak a variety of languages, their clamor is compared with the bleating of innumerable ewes waiting to be milked in the courtyard of a rich man (4.433–37). Later, they are briefly in danger of being penned inside Troy like lambs by Diomedes, who charges at them on his chariot (8.130–31). In a new permutation, somewhat later, a human attacks other humans in a way that is compared to a man aided by larger animals: one of the most flamboyant similes in the poem compares Ajax, striding up and down the decks of the ships with a long pike, urging on his comrades with the vigor of a man who can leap between the backs of four horses that are harnessed together, charging across a plain at speed (15.675–86).

Sometimes different types of the same animal are used to describe the relationship between humans, as when Menelaus stands over the corpse of Patroclus to defend it from the Trojans, like a young cow lowing over her first newborn calf (17.4–8). Some similes with threefold comparands can mix both categories. When Aeneas looks with pleasure on other Trojan captains leading the Trojan army—Deiphobus, Paris, and Agenor—he is compared with a shepherd watching sheep follow a ram from their pasture-land to water (13.489–95).

The terms of comparison shift constantly, creating confusion as to the precise significance of humanity in relation to the animals it domesticates: are all humans like animals, or does inferior social class make a human like an animal under the power of its pack, flock, or herd leader? The permutations are seemingly (almost) infinite: the world of Homeric analogy can never make up its mind.

Man vs. Wild Fauna vs. Domesticated Animals

The agents, collaborators, subjects, and victims in the poem's narrative and similes may be wild fauna on land and in the air and sea as well as domesticated fauna—in addition to gods, elemental or meteorological phenomena represented by gods or not, humans of all social classes, and even, sometimes, inanimate objects such as weapons. But more numerous similes involve humans attempting to kill or catch wild creatures, either to eat them or to ward them off from their domesticated livestock. The constant chaotic shifts in agency imply a world in which the work of agriculture and of acquiring human and animal shelter—hunting and fishing, the provision of food, products made from animals such as leather—is fraught with menace and disharmony. The situation is best summed up by Achilles in his grim statement to Hector that they can neither be friends nor make oaths between them until one of them has glutted Ares, the warrior with tough shield of hide; the conflict between them is as embedded in nature as the enmity both between wild lions and men and between wild wolves and lambs (22.261–67).

The *Iliad* focuses far less on the sea than does the *Odyssey;* the presentation of meat, especially beef, as the source of protein best suited to warriors, results in a far smaller proportion of images related to fishing than to hunting on land. The fourth-century Athenian comic poet Eubulus made a character ask, "Where does Homer ever speak of any of the Achaeans eating fish?"[45] Ancient commentators, while univocally noting the absence of fish from the diet of the *Iliad*, could not decide whether it was because fish was regarded as a lower-class nutriment, and cooking in the utensils necessary for successful seafood cuisine was too humble a task for heroes to perform, or because fish was regarded as a luxury liable to effeminize its consumers.[46] Yet even here there are exceptions.

Patroclus impales Thestor through the jaw, and drags his corpse over the chariot rim, like a man sitting on a rock who drags a "sacred" fish out of the sea with a fishing line and bronze hook (16.406–8). Iris plunges down to the sea in search of Thetis like a lead plummet, fastened on an ox's horn, sent down by fishermen to ensnare ravenous fish (24.80–82), a simile recalled by Plato's Socrates when he is proving to the rhapsode Ion that a fisherman will know more about fishing than a bard.[47] But even fish can become predators, at least on dead bodies. Agamemnon sacrifices a boar "with pitiless bronze" to underline the oath claiming that he never had sex with Briseis, and Talthybius throws the body into the sea "to be food for fishes" (19.266–68). Achilles hurls Lycaon's corpse into the River Scamander, and boasts that instead of receiving a mother's lamentation, the corpse will be subjected to pitiless fish licking the blood off its wounds (21.123–25).[48] Fish can also be the prey of larger sea creatures, as in the simile where Trojans in the Scamander fleeing Achilles are compared with fish trying to escape an enormous dolphin (21.17–26).

The correspondences can become complicated and far from unilateral. A hero may be like a wild animal beset by both humans and other wild animals, even fauna of more than one type. In a complex simile, Odysseus is surrounded, like a stag that has been wounded by a hunting bowman in the mountains and encircled by a pack of jackals. But the jackals are dispersed by a lion, who makes the stag its own and eats it. A human wanted to eat the stag; so did both jackals and lions, but the true victor of the hunt is the lion, who represents the mighty Ajax, intervening and warding them off (11.472–88). Humans are in this simile seen as just one of three types of predatory mammal, and hardly the most successful.

In by far the most frequent of the similes involving wild animals, however, they predate, or attempt to predate, on animals that humans have domesticated. This introduces human habitations into the picture. Such similes, by their comparative function, set up humans as if they are in direct and often life-threatening, violent, territorial competition with untamed fauna, as well as in positions of being lords and masters over their domesticated beasts. There are triangular scenes featuring wild animals attacking livestock, not in the hills or plains or marshes, but in or beside a sheepfold or farmstead, a manmade construction housing domesticated livestock. Such scenes underpin several similes. Diomedes, during his series of notable feats on the battlefield in book 5, is compared in two of these to a lion that attacks domesticated animals. Fury enters Diomedes's heart in three waves, as it does in a lion that a shepherd has succeeded in wounding but not killing as it jumped over the wall of a sheepfold. This maddens the lion to greater violence, so it drives the sheep in chaotic flight and leaves their corpses in heaps, before leaping out over the wall once more (5.136–42). A thoroughly unfruitful occasion, for both beasts and man.

Woods, pastureland, and mountainsides, even where no human building has been constructed, are also disputed territories that lions roam in their quest for food. Diomedes kills two sons of Priam (5.161–62) like a lion coming across a cow grazing in the woods, and breaking its neck. In the "Doloneia," he subsequently kills the sleeping Thracian warriors, as a lion springs on a flock of sheep or goats when their herdsman has left them unattended, and they groan terribly as it hacks at them (10.484–88). Agamemnon chases the Trojans over the plain like a lion who has attacked cows in the middle of the night, grabbing one in its jaws and gorging himself on its innards (11.172–76).

The Achaeans are driven in rout by Hector and the Trojans, like a herd of cattle or a great flock of sheep being driven in confusion by two wild beasts under cover of darkness (15.323–26). Hector attacks the enemy like a malevolent lion leaping on myriad cows as they graze. They are under the supervision of a novice herdsman who does not know how to fight over a carcass, and who also tends only the front and the rear of the herd, neglecting the middle of the column (15.630–37).

When the Greek leaders do rally, they fall on the Trojans like wolves on lambs and kids, separated by an ignorant herdsman in the mountains from the adults of the flocks (16.352–54). When Patroclus smites Sarpedon in the midst of his Lycians, he is compared with a lion who kills a proud bull in the midst of the herd (16.485–90). And later, the Greek Automedon, Achilles's charioteer, after killing Aretus in revenge for Patroclus's death, strips him of his arms, places them on the chariot, and rides off, his hands and feet steeped in gore "like a lion that has devoured a bull" (17.540–42). Hector swoops on Achilles like a soaring eagle darting down to the plain to seize a lamb or hare (22.308–10).

There is even one elaborate double simile, which reveals the poet's perhaps unconscious apprehension of an analogy between man's ability to destroy wild animals and wild trees. The Achaean brothers Crethon and Orsilochus killed by Aeneas are like two lion cubs raised in the forest. They snatch cows and sheep and create havoc in men's farmsteads, until they are killed by the hands of men with sharp bronze and fall like tall fir trees (5.554–60). The sharp bronze wielded by humans threatens human enemies, but it threatens animals and forests as well.

Dogs on the Threshold

Many of the similes describing wild animals preying on domesticated ones feature dogs working in collaboration with humans to preserve the livestock from which they both benefit. These are dogs that watch the herd but that can also get involved in the chase if a predatory wild animal is encroaching. In similes involving an agricultural construction, the dogs are seemingly domesticated enough to cooperate with peasants who are defending it from a predatory wild animal. The Achaean watchmen listen out for the Trojans as hard as dogs guarding a sheepfold (10.183–85). Ajax retreats before the Trojans like a lion that is attempting to spring into a stockyard, but is repelled by dogs and peasants (11.549–54). Sarpedon attacks the ships like a mountain lion driven by hunger to jump into a sheepfold, only to encounter shepherds guarding the flock "with dogs and spears" (12.300–306).

The two Ajaxes carry Imbrius away like two lions that have snatched a goat from sharp-toothed dogs (13.197–200). Menelaus leaps on Euphorbus to strip him of his armor like a mountain lion that has seized a domesticated heifer, breaking its neck in his jaws and gorging on the heifer's blood and guts while hounds and herdsmen clamor around (17.61–67). Yet not long afterward, Menelaus is forced to retreat from Patroclus's corpse like a bearded lion that "dogs and men" successfully drive away from a sheepfold (17.110–11). Still later, he departs from the fray like a lion that tries to seize the fattest of a herd of cattle, but grows tired of doing battle against the dogs and men guarding it and retreats under the onslaught of spears and flaming brands (17.656–64).

There are, however, several similes in which dogs are primarily used by men to hunt wild animals rather than guard domesticated livestock. Here the dogs, although still cooperating with humans, are behaving as they would in the wild. Menelaus is compared with a lion that alights on the carcass of a stag or wild goat, and succeeds in eating it despite being set upon by dogs and youths (3.23–26). Hector snaps at the heels of the Achaeans like a hound getting his teeth into the buttock or flank of a fleeing wild boar or lion (8.338–40). Odysseus and Diomedes chase Dolon like two sharp-toothed hounds running down a doe or hare in the woods (10.360–62).

Hector also sets his Trojans upon the Achaeans like a hunter setting his white-toothed hounds upon a wild boar or lion (11.291–96), but here the human agent is using the way that dogs in the wild kill for food to his own advantage, thus erasing the binary opposition of wildness and tameness. The Trojans press on Odysseus like dogs and youths charging on a wild boar coming out of a thicket (11.414–18).

The Lapiths fight in front of the Achaean gate like wild boars holding their own in the face of a throng of men and dogs attacking them (12.145–49). Idomeneus withstands the Trojans like a boar warding off dogs and men "in some isolated place," back bristling, eyes blazing, and tusks menacing them (13.471–75). The Achaeans press on but are dismayed by the sight of Hector; they are like dogs and peasants pursuing a stag or wild goat, who mislay sight of their prey and then lose it to a lion (15.271–76). Ajax scatters Trojans by striding straight through the front ranks like a wild mountain boar seeing off hounds and youths (17.281–83). The Trojans battling for Patroclus's corpse charge like hounds who lead youths to a wounded boar, but then are beaten back and give ground (17.725–29). Agenor refuses to flee from Achilles like a leopardess that holds her ground when assaulted by a hunter, unafraid of the baying of hounds, and is determined to fight to the death despite being pierced by a spear (21.574–79).

Dog symbolism becomes a small but significant part of the imagery in the buildup to the final showdown between Hector and Achilles. When dying, Hector begs Achilles to spare his parents by not abandoning his corpse to be gorged on by dogs beside the Achaean ships (22.338–39): the ruthless victor retorts, "Do not beg me, dog, by knees or parents" (22.345). But the imagery has previously associated Achilles in this battle, rather than Hector, with a dangerous lone dog, for once unaccompanied by humans: Priam watched Achilles speeding over the plain like the brilliant star that men call the "dog of Orion," a sign of evil that brings fever with it (22.29–31).[49] Achilles pursues Hector like a hound that chases a fawn out of his covert; the fawn may escape temporarily to cower beneath a thicket, but the hound will relentlessly track it down until he catches it (22.189–92).

As James Redfield saw so clearly, the most significant animal at the *Iliad*'s fuzzy boundary between "nature" and "culture"—its "natureculture" according to Holmes—is the dog.[50] In the main narrative of the *Iliad*, dogs are normally hunters or watchdogs.[51] Yet the role of these and other dogs is ambivalent. They can cooperate with humans, but their communities are somehow liminal. They lurk on the physical thresholds of human communities, undermining the binary conceptual opposition not only between nature and culture, but also between "wild" and "domesticated" animals.

Hera recalls Heracles bringing the unnamed hound of Hades, elsewhere called Cerberus, out of Erebus (8.368), but this is the nearest the *Iliad* comes to an individual pet dog equivalent to Argos, Odysseus's loyal hound, in the *Odyssey* (17.290–319).[52] Nor is there an equivalent of the death of Actaeon, torn to pieces alive by his own hunting dogs when they turned on him, although Priam comes close to describing such a scenario when he fears being eaten at his own front entrance by the very dogs he had fed from his table so that they would guard his door (22.66–71).[53] Patroclus had nine dogs with whom he shared his tables (23.173), and two of them are sacrificed, along with horses and young Trojan men, at his funeral (23.170–76).[54]

But the opening sentence of the epic announces that it will sing of Achilles's wrath, which caused the corpses of many heroes to become prizes for dogs and birds (1.1–5); the implication is that famished stray dogs roam the battle plain, parasitically waiting for carrion. This, however, is the sole instance of carrion dogs appearing in the main narrative, in the authorial voice, rather than in threats or as yet unfulfilled wishes or anxieties.[55] It has been suggested that these wild carrion dogs were the Trojans' dogs, "driven from their natural homes and forced by semi-starvation to the work of scavengers."[56] Athena bloodthirstily relishes the prospect of many Trojans glutting the appetite of dogs and birds with their fat and flesh when they have fallen at the Achaean ships (8.379–80). Hector gleefully imagines Ajax's fat and flesh glutting the Trojans' dogs and the birds (13.831–33).

Menelaus and Iris both fear that Patroclus's corpse, if not retrieved from the battlefield, will become the "sung entertainment" of the dogs of Troy (17.255, 18.179), an extraordinary term that almost suggests dogs listening to a recital of poetry as they feast on human flesh.[57] Zeus does not want Patroclus to be the "booty" of dogs (17.272).[58] Athena warns Menelaus that Patroclus's cadaver is at risk of being "torn by swift dogs beneath the Trojans' wall" (17.558). Priam wishes that the gods felt the same about Achilles as he does, for then Achilles would soon lie prostrate, being eaten by dogs and vultures (23.41–43). Aphrodite keeps the dogs off the corpse of Hector by anointing it with rose-sweet ambrosial oil (23.185–87).

Dogs, moreover, feature unattractively several times in dyspeptic rhet-

oric that accuses others of shamelessness, unfair division of spoils, or failure
to live up to responsibilities: the scholiasts routinely gloss the metaphori-
cal use of the term "dog" with the adjective *anaidēs*, "shameless."[59] Achilles
calls Agamemnon "dog-faced" when he is accusing him of exploiting the
Achaean soldiery (1.159); he later says that Agamemnon is welcome shame-
lessly to deceive all the other Achaeans, but that he would not dare to look
Achilles in the face despite his own face being that of a dog (9.370–73).
Hephaestus says his "dog-faced" mother, Hera, cast him out of heaven
because she disliked his lameness (18.396); Helen refers to herself as dog-
faced in the contexts of her abandonment of Menelaus (3.180) and the
trouble she has caused everyone (6.344). Both Diomedes and Achilles address Hector as "dog" when he is rescued by Apollo (11.362; 20.499). The
abusive term "dog-fly," applied to Ares by Athena and to Aphrodite by Hera
(21.394, 21.421), may refer to an insect, but the aural effect of the canine
first half of the word deepens the insult.[60] When Menelaus explicitly accuses the Trojans of offending Zeus Xenios, the god who oversees the
safety of guests in alien communities and hosts receiving alien guests, he
calls them "evil bitches" (13.623).[61]

The *Iliad*'s dogs are the poem's master symbol of its unstable boundaries between legitimate and shameless violence, between tame and wild,
between civilized and bestial. Like Ares, the god who will fight on either
side (5.831, 5.889), dogs are always potential renegades: given the opportunity, they will turn against the very humans who have nourished them
and devour their corpses. They are both inferior to humans, serving them
as guard and hunting dogs, and, if not controlled correctly, potentially
inimical. Their social, moral, and ontological hybridity marks the very
center of the poem's ideological world, which teeters on the brink of collapsing into fearful anarchy.

CHAPTER SEVEN

Smiths

Automedon spoke, poised his far-shadowing spear and cast it,
striking Aretus' shield, which was well-balanced on either side.
But it did not halt the spear, for the bronze went all the way through,
and he rammed it through Aretus' belt into his lower belly.
It was as when a strong man, sharp axe in hand,
strikes an ox from the pasture behind his horns,
to cut right through, and the ox springs forward before collapsing.
That was how Aretus sprang forward and fell flat on his back,
and the sharp sword quivered as it stuck in his bowels, loosening his
    limbs.

In this unusual simile from the *Iliad*, the slaughter of a sacrificial animal with a metal implement is likened to the death of Aretus by spear on the battlefield. But the analogy between animal sacrifice and violence committed by humans against humans, an analogy that was to become such a prominent feature of Athenian tragedy, is otherwise alien to the poem, except perhaps in Diomedes's reference to Lycurgus striking down with an ox goad the maenads who had nursed Dionysus (6.130–35).[1]

Instead, it is with the harvest that Odysseus compares battlefield slaughter when he delivers his remarkable warning to Achilles that men tire of battle when the loss of life is out of all proportion to the gains being made (19.221–23):

> Humans speedily feel sated with the din of battle
> when the bronze spreads most straw on the ground
> but the harvest is most meagre.

It has been argued that the "harvest" here is not enemy corpses but the loot in the form of arms that the victors strip from the dead, or from the ground, when the enemy has fled in rout. It is a typical feature of battle-field conventions in a poem where elite male status relative to peers is partly determined by the quantity of such goods he amasses.[2] Yet, in either interpretation, "the bronze" here is in the nominative case and is invested with a deadly agency of its own. There are other instances. "The bronze" does not break through Menelaus's shield (3.348).[3] "Spears hurled by bold hands" become lodged in Ajax's shield "as they speed onwards": "Many, before they reached his white flesh, fixed themselves in earth, yearning to taste his flesh" (11.570–74; see also 15.314–17 and 21.167–68). Bronze deprives sacrificial lambs of life (3.294). The bronze penetrates deep into a lung (4.528, see also 20.486). When Agamemnon kills Hyperenor, "the bronze loosened his bowels as it drove through" (14.516–18, see also 17.345–46). "The bronze passed right through, and he drove it through the belt into the lower stomach" (17.518–19). "The bronze leapt back from the man it struck and did not pierce him through" (21.593–94).[4]

Weapons that may be assumed to have bronze components, especially spears, are also depicted in the act of taking a life and thirsting for blood.[5] In the *Odyssey*, iron weaponry is given similar agency: it does not just scythe men down like straw, but by its mere presence encourages them to fight. Odysseus tells Telemachus how to explain to the suitors why he is remov-ing all the palace's weapons of war (16.284). He is to say that he fears that the suitors, when drunk, will quarrel and wound one another, "for the iron itself drags a man to it" (16.294). Telemachus later repeats these words verbatim to the suitors (19.13).[6]

This proverbial expression assumes that the mere availability of metal weapons incites men to violence. Just five pithy words encapsulate the epoch-making truth that war, of the nature and on the scale depicted in the *Iliad*, only became possible with the substantial advances in metallurgy that produced the Bronze and Iron Ages in the first place. Propelling stone missiles or wood and bone darts at a foe does not make siege warfare possible. Nor are ceramics suitable for warfare: as Theodore Wertime writes, "Metallurgy represented the first serious penetration of the earth by men and fire. While ceramics offered economically more important pyrotechnic products at first, metallurgy introduced men directly to the science of the earth and its materials. It also lent itself directly to war, in a way that ceramics could not."[7]

The proverb may have entered the Homeric poets' repertoire at a relatively late stage, since the Greeks started using iron on a large scale only in the twelfth or eleventh century BCE, much later than bronze. But some of the language used to describe metal objects may belong to the poem's early, Mycenaean, stratum of Greek. The silver-studded sword that Agamemnon throws around his shoulders after being visited by the dream "is probably a very old formula since its three component terms are all found in Linear B. Also such swords, with bronze blade joined to the handle by silver rivets, are Mycenaean."[8] The dative *chalkophi* in the phrase "bronze was turned aside by bronze" (11.351) must be of Mycenaean origin.[9]

Although to modern audiences the scale of bronze manufacture envisaged in the *Iliad* may seem like wanton and wildly excessive consumption of nonrenewable natural resources, there is no archaeological evidence to suggest that the high-status Homeric gods' and warriors' profusion of metal objects would have seemed anything but admirable to most eighth-century listeners to the poem. Archaeologists have found at Olympia a flood of eighth-century dedications of high-quality bronze tripods, "an ideal status symbol and commodity for conspicuous consumption," suggesting "a material emphasis on display rather than communality."[10] The *Iliad* is replete with fantastically forged and invaluable metal objects like the divine aegis, decorated with a hundred golden tassels, all cunningly twisted, "and each one worth a hundred oxen" (2.448–49). It shines with terrifying brightness; Hephaestus had made it for Zeus (15.309–10). Agamemnon tries to bribe Achilles to return to the fighting with seven tripods, ten talents—a stupendous quantity—of gold and twenty gleaming cauldrons as well as seven beautiful slave women (9.122–24); Achilles eventually accepts them (19.247).[11] Nestor's tableware includes a basket made of bronze, and a four-handled cup too heavy "for men of today" even to lift, studded with golden bosses and images of doves (11.631–37).

We have seen how deforestation was accelerated by logging for construction, arms, ships, land clearance for pasture, and firewood for cooking, heat, and light. But just as significant in causing deforestation was metal-

lurgy, especially the smelting and forging of bronze and iron on an industrial scale.[12] The fuel needed to smelt iron was charcoal, that is, wood
baked (not burned) to extract impurities, leaving only carbon.[13] Between
20 and 28 million hectares of trees were required to smelt enough to leave
behind the 50 to 90 million tons of Iron Age slag found round the Mediterranean; the iron smelters of Populonia alone are said to have consumed
on average 375 hectares of wood, or the equivalent of 926 soccer fields,
annually.[14]

A detailed study of industrial waste, including charcoal, in the Timna
Valley within Israel's southern desert area during the eleventh to ninth
centuries BCE, offers what must be a typical and representative example
of the destruction of trees in the cause of metal production in the eastern
Mediterranean.[15] The region, today a desert, was at the time the location
of thousands of mining sites and around ten processing centers, where
copper was extracted from ore in high-temperature furnaces. From the
scale of the remains, the Timna researchers have calculated the quantity
of woody plants required for producing copper. At a single production site
called "Slaves' Hill," they believe that as many as four hundred acacias and
1,800 brooms were incinerated every year. In 80 percent of the early samples, the fuel came from local trees, acacia thorn and white broom. The
Bible even refers to the excellent firewood provided by "burning coals of
the broom tree."[16]

But the charcoal fuels that were used in copper production changed
perceptibly over time. By the mid-tenth century BCE, as these resources
dwindled, the industry looked for other solutions, burning lower-quality
fuel from desert bushes and palm trees. When those plants ran out, juni-
pers began to be imported, from as far away as a hundred kilometers, from
the Edomite plateau in Jordan. Transporting woody plants from afar did
not prove economically viable in the long run, however, and eventually,
during the ninth century BCE, all production sites were shut down. The
industry was conducted unsustainably and therefore failed, concurrent with
the disappearance of vegetation; the researchers say that the local environment has not recovered fully even today.[17]

Iron is used everywhere in the *Iliad* for everyday tools: knives, axes, and
all the implements of the domestic and agricultural life that receive such
minimal attention in the main action of the poem relative to the activities
of warriors. But a nameless man in a simile chops down a tree with an iron
implement (4.485). Pandarus's arrows have iron heads (4.123). Nestor recalls King Areithoüs, whose unique weapon of choice was an iron club or

mace, with which he broke up battalions (7.141). The gates of Tartaros that loom over its bronze threshold are made of iron (8.15). Antilochus fears that the bereaved Achilles may cut his throat with an iron knife (18.34). An iron knife is used to sacrifice the livestock at the funeral games (23.30).

The prizes that Achilles brings from his ships to be awarded at the funeral games include cauldrons, tripods, horses, mules, oxen, women, and gray iron (23.259–61).[18] The prize awarded in the field event that seems to resemble throwing the discus or hammer, or putting the shot, doubles as the object that the contestants need to hurl as far as possible—a solid mass of rough-cast iron (23.826).[19] Achilles claims that it would provide the winner's shepherd or plowman with enough iron to come to the city for five years (23.833–35). This somewhat obscure expression may mean that Achilles assumes peasants forge their own iron implements at home, going to town to purchase metal. But it certainly implies the strength of the contestants, since the mass is of huge size, and could provide material for numerous items of shearing equipment or parts of a plow: the men's hammer in modern athletics, which weighs sixteen pounds, is not remotely equivalent. The first and second prizes in the archery contest are ten iron double axes and ten iron single axes, respectively (23.850–51).

Achilles says he has already accumulated substantial loot to take home from Troy, "gold and ruddy bronze and fair-girdled women and grey iron" (9.365–66).[20] Dolon is a man "rich in gold and rich in bronze" (10.315–16), and can offer to ransom himself. He tells his captors that his father can supply an "immeasurable ransom" because his household contains "bronze and gold and iron that requires much labour" (10.378–80).[21] Ownership of "bronze, gold, and iron which requires much labour" recurs as an epic formula for describing a wealthy man (11.132–33). When Adrastus begs Menelaus for his life, he too says he can offer an "infinite" ransom, for his father has great treasure stores: "bronze and gold and iron that requires much labour" (6.48). The epithet "that requires much labour," *polukmētos*, is derived from *polu-* ("much") and the verb *kamnein*, which, as so often in Greek thought, elides the idea of doing physical work and toil with the endurance of misery and distress. Yet that intensive labor, toil, and distress are scarcely visible in the human world depicted in the *Iliad*. The substance constituting every single metal object in the poem must be assumed by the listener to have already passed through the hands of several sweating laborers in mountains, mines, transport vehicles, furnaces, smithies, and forges.

By the late Bronze Age, little metal of any kind came in chunks or lumps lying in obvious, convenient locations in creek beds, in rocks, or visibly on

cliff faces. It was humans who had to dig for their ore, even if it was the
Aeschylean Titan Prometheus who claimed to have been the first to discover the "benefits" that they accrued from the bronze, iron, silver, and
gold "hidden beneath the earth."[22] First they dug open-cast trenches and
shallow quarries, then they crawled ever deeper into the earth.[23] Mineshafts twenty meters deep were commonplace; they were created by fire
setting and pounding with hammers and mortars made of both stone and
bronze. Mining was even more arduous than mineshaft-setting, and was
dark and dangerous; miners were vulnerable to suffocation by collapsing
shaft walls and especially to flooding.[24] Many Bronze Age galleries are low,
intricate, and narrow; miners would have often been obliged to lie on their
backs in cramped conditions and unfree child labor must be assumed.[25] It
has been estimated that progress was so slow, since miners could dislodge
only small chunks of rock at a time, that they cannot have advanced more
than twelve centimeters in the course of a ten-hour day.[26] Miners used
equipment that had already cost hundreds of hours of human labor—pick-
axes, hammers, chisels, buckets, baskets, ropes, and wooden scaffolding. Yet
however hard they worked, it was always for diminishing returns. Ore veins
in the later Bronze Age became thinner and more fragmented and yielded
far less usable raw material.

Bronze is an alloy of 90 percent copper and 10 percent tin. But raw
ore hacked out of other rock is rarely pure; it can have veins of other minerals coursing through it, or appear only as fragments implanted in other
rocks. The isolation of pure gold, silver, copper, tin, or lead requires a
complicated process of smelting in a furnace at blisteringly high temperatures. Copper drips out of surrounding rock only at two thousand degrees
Fahrenheit.[27] The wood charcoal required to attain this high temperature
was manufactured by specialist charcoal-burners at dedicated sites, by the
slow combustion of trees and bushes cut down for this purpose.[28] Sometimes furnaces were built adjacent to mines, but often there was an inter-
mediate task to be performed, since the raw rocks were transported from
mineshafts to workshops at considerable distances apart.

The soft smelted ore was then cast into molds, originally of stone, and
later of clay hard enough to withstand the heat of the molten metal. Molds
were developed that consisted of two facing hollow parts, with clay cores if
a hollow object such as a spearhead was required. Bronze molds were also
used for bronze in a process called "gravity casting." Working with molten
ore at high temperatures was appallingly dangerous, with the workers at
constant risk of lethal burns. Some items required that the cast metal, once
released from its mold, be hammered and further crafted intensively. Sword

edges were hammered to wafer-thin sharpness; sheet metal was produced by intensive human labor at the anvil. Men applied hammers to crude ingots that had merely been smelted and cast. The metal often needed to be reheated repeatedly in the process to anneal it, since otherwise it would become too hard and friable.[29]

The *Iliad* conceals almost every stage in this extraordinary process of extraction and production, since the unforgettable portrait of Hephaestus at his forge in book 18 picks up only at the point where he has at hand ingots and sheets of processed metal. For every shield, helmet, breastplate, greave, and weapon that clangs as it is worn by a single lauded warrior in the *Iliad*, hundreds of hours of human labor must be imagined to have been expended in dangerous mines and workshops by innumerable workers whose names and personalities are lost to history and literature.

Just one entry in the Trojan catalogue perhaps prompts the listener to imagine a mine: on the Trojan side, the Halizones were led by two men named Hodios (later killed by Agamemnon at 5.39) and Epistrophos; they came "from distant Alybe, which is the birthplace of silver" (2.856–57).[30] Halizones may mean "Sea-Belted Ones," suggesting that they came from a distant part of Pontus, but it has been suggested that the name may recall the River Halys (now Kızılırmak), which was a Hittite name; the Hittites were major suppliers of silver to the Greek-speaking world in the second millennium BCE.[31]

Divine and Elite Gold

The incessant clang of metal in the *Iliad* begins with Apollo storming down from Olympus to shoot plague-arrows at the Achaeans: "the arrows rattled on the shoulders of the wrathful god as he moved, and he came like the night . . . the twang of his silver bow was terrible" (1.46–49). The gods, presumably since they are projections of the richest and most powerful mortals in a monarchical society, enjoy fabulous metal artifacts, especially golden and silver ones.[32] The floor of Olympus is golden (4.2) and the gods drink nectar from golden goblets (4.3). Their thrones—notably those of Hera, Artemis, and Dawn—are made of gold (1.611, 8.436, 8.442), as are even the clouds surrounding the mountaintops of Olympus (13.523) and those with which Zeus can surround the summit of Ida (14.343–44). Athena, too, has access to golden clouds, for she can cast one around Achilles's head when she flings the aegis around his shoulders (18.203–6).

The aegis is probably the single most valuable object in the poem, with its hundred pure gold tassels, each worth a hecatomb of cattle (2.448–49). Ares's horses wear golden frontlets or muzzles on their faces (5.358); Aphrodite and Hera both wear golden brooches (5.425, 14.180); Apollo's sword is made of gold (5.509); Artemis's reins are gold (6.205), as are her arrows (16.183). Hephaestus has put golden wheels under his self-moving tripods so that they can shift themselves automatically between his house and the hall of all the gods (18.375); the robotic handmaidens who attend the lame god are likewise made of gold (18.417–21).

Hera's chariot is one of the most elaborately crafted objects in the poem. Her horses, too, have golden frontlets and breast straps (5.720, 5.730–31), while the wheels and tires are bronze, the axle-tree iron, the fellow gold, the naves silver, the car made of plaited strips of gold and silver, the pole silver, and the yoke itself is gold (5.722–30). Hera bribes Sleep with the offer of an imperishable golden throne made skillfully by her own son Hephaestus, along with a footstool (14.238–41). The wings of her henchwoman Iris are golden (8.398, 22.209). Zeus reminds Hera that he had once punished her by hanging her up, with anvils suspended from her feet and her wrists tied together with an unbreakable band of gold (15.18–21).

Zeus has also challenged the other gods to try to drag him down from heaven by a golden chain (8.19); shortly afterward, as if to reinforce the point, he departs on his horse-drawn chariot in a blaze of gold, mentioned three times in three consecutive lines (8.41–45):

> So saying, he had his bronze-hooved horses harnessed to his chariot;
> the pair are fast in flight, with flowing golden manes.
> In gold he arrayed his own body,
> golden was the well-made whip he grasped.

The scales on which Zeus weighs warriors' souls are likewise golden (8.69).

Poseidon's underwater palace at Aegae is "golden and gleaming" and "imperishable forever" (13.22). The same formula used to describe Zeus embarking on his chariot in a blaze of gold at 8.41–45, quoted earlier, is repeated for Poseidon here (13.23–26); when he releases his horses from the chariot to graze, he places unbreakable golden fetters around their legs to prevent them from wandering away (13.36).

Humans use gold and silver too, but more sparingly. Hector's enormously long spear has a bronze point encircled by a ring of gold (6.320). The normal loot taken from small towns around the Troad is bronze in the form of bronze ingots and arms, as well as women; gold is more likely to be housed in Troy and used in costly ransoms.[33] Thersites complains that Agamemnon has already acquired plentiful bronze and women; is it only gold that will suffice him, he demands to know, exchanged in a ransom for some Trojan's son whom Thersites or some other soldier has been the one to capture (2.229–31)? But important symbolic and ritual objects in the world of men are often golden—Chryses's priestly scepter (1.374), crowned with Apollo's wreaths, or the golden scepter with which Odysseus raises weals on Thersites's back (2.268). At Patroclus's funeral, Achilles pours libations to the winds, asking them to help kindle the pyre, from a golden cup (23.196), and pours wine on the ground for his dead comrade from a golden bowl (23.219); the bones are gathered in a golden urn (23.253) with two handles, which Thetis had bestowed on her son (23.91–92). The first prize in the footrace at Patroclus's funeral games is the dead man's own huge silver mixing bowl, the loveliest in all the world, crafted by skilled Sidonians and transported across the sea by Phoenicians (23.740–47).[34]

It seems to be a sign of ostentation when the Trojans' allies arrive for battle equipped with gold, preferring it to bronze. A Carian leader arrives for war "all decked with gold, like a girl" (2.872–75), but it did him no good, since he was later killed in the river by Achilles. Glaucus affirms his family's ancient guest-friendship with Diomedes by recalling that Bellerophon had once bestowed on his father a double cup of gold (6.220). Glaucus, however, seems to wear golden armor, which he naively exchanges at great personal loss for Diomedes's bronze arms, "the value of a hundred oxen for the value of nine" (6.236). King Rhesus has come from Thrace with a chariot made of silver and gold and a fabled suit of golden armor of exquisite workmanship, which even Dolon, ostensibly on the same side as Rhesus, says is too magnificent for a mortal and suitable only for gods (10.338–41).[35] The Trojan Antimachus believes that if he argues against returning Helen to the Achaeans, Paris will give him rich gifts of gold, as a reward (11.132–35).

Yet there are ostentatious arms and armor on the Achaean side as well. In one of the most elaborate, gorgeous, visually arresting descriptions of metalwork found in the *Iliad*, Agamemnon arms himself in exceptionally valuable armaments decorated with metals of contrasting colors (11.17–46):

> First, he put his greaves on his shins;
> they were beautiful and fitted with silver ankle-pieces.
> Second, he fitted his breastplate round his chest,

which Cinyras had once given him as a token of guest-friendship.
He had learned in Cyprus the momentous report
that the Achaeans were about to set sail in their ships for Troy.
That's why he bestowed it upon the king to please him.
There were ten bands of dark enamel on it,
twelve of gold and twenty of tin.
Dark enamel serpents coiled up towards the neck,
three on either side, like rainbows which the son of Cronus
plants in the cloud as a portent for men endowed with speech.
And he threw his sword around his shoulders, the golden studs
gleaming, while the scabbard around it
was silver and fitted with golden chains.
He picked up the man-sized richly wrought shield,
fit to run into battle with and beautiful, with ten bronze roundels,
and on it there were twenty knobs of pale tin,
with one in the middle of dark enamel,
and it was crowned with a Gorgon of grim appearance,
glaring terribly, and around her were Terror and Fear;
the baldric was made of silver. And on it
a dark enamel serpent thrashed; it had three heads,
pointing in different directions, growing from a single neck.
On his head he put the helmet with two horns, four bosses,
and a horsehair crest; the plume nodded terrifyingly above.
Then he grasped two doughty spears, tipped with bronze,
and sharp. The bronze gleamed from them all the way to heaven.
Athena and Hera applauded noisily,
honouring the king of Mycenae, rich in gold.

Mycenae is also "rich in gold" at 7.180. In the world of the poem, there is scarcely any consciousness that war depletes the wealth in metal goods owned by such fabulously wealthy kings.

At just three moments of forthrightness, Achilles and Hector, on the topic of accumulated wealth, both sound more like Hesiod than Homer. Achilles says that not even all the wealth that people say Troy *formerly* possessed, in a time of peace, is worth as much as life to him (9.401–3). Hector tells his "unnumbered" allies that he did not gather them in such large numbers for the sake of it, but in order to save the Trojans' wives and children; that is why he is wasting his own people's stores by giving the allies food and gifts so that their strength flourishes (17.221–26). Later,

Hector rebukes Polydamas for suggesting that the Trojans retreat inside the walls. Has he not had enough of defending the city against the Achaeans (18.288–92)?

> People used to tell of Priam's city,
> and its abundant gold, abundant bronze.
> But now its lovely treasures are completely wasted,
> and many of its possessions sold away to Phrygia and fair Maeonia,
> since the wrath of great Zeus began.

Supplies of metal are not infinite. Even the wealthiest city can run out of wealth. It is just that Homeric warriors rarely admit it.

The Sensory Assault of Bronze

Gods and the richest and most powerful men, the likes of Agamemnon and Priam, possess gold; tools and two weapons are made at least in part of iron; to make all the bronze items, tin must be assumed to have been combined, in a small ratio, with copper; tin also appears on greaves and in decorations on other pieces of armor. A "greave of newly worked tin rang terribly" on Achilles (21.558–59).[36] But the metal that eclipses all others in the *Iliad* is undoubtedly bronze. Bronze is the metal from which arms were made, for it could be tempered far better than iron. The technology to make iron swords practicable was not used by Greeks on any scale until the late eighth century BCE. Iron weapons bend and break far more easily; Polybius describes the shockingly poor quality of the iron swords wielded by the Celts who invaded Italy in 225 BCE: "they could give only one downward cut with any effect, but afterwards the edges got so turned and the blade so bent, that unless they had time to straighten them with their foot against the ground, they could not deliver a second blow."[37] A tool such as a woodcutter's axe can support a heavy weight and mass of iron behind its sharp edge in a way that a sword or sword blade cannot.[38] But even Homer's bronze weapons break: swords crack at the hilt when they strike shields or helmets: Menelaus's sword breaks into pieces when he smites Paris's helmet ridge; Lycon's sword breaks off at the hilt when he hits Peneleos on the socket of his helmet crest (3.367, 16.339).

There are nearly 450 instances of words related to or compounded with "bronze" (*chalkos*), besides the ubiquitous adjectival *chalkeos*, which is liberally applied to thresholds, mansions, parts of vehicles, spears, swords, shields, breastplates, and helmets (for example, 8.15, 18.371, 13.30, 5.723, 3.317, 3.335, 7.220, 13.398, 13.440, and 18.131). These related words include "work in bronze" (18.400), "bronze-smith" (4.187, 4.216, 12.295), "fitted with bronze" (3.316, 4.469, 5.145, 13.650, 15.535, 15.544, 17.268), "with bronze breastplate" (4.448, 8.62), "heavy with bronze" (15.465, 11.96), "bronze-tipped" (22.225), "of fine bronze" (20.322), "all of bronze" (20.102), "abounding in bronze" (5.504, 10.315, 18.289), "armed with bronze" (5.699, 6.199, 6.398), "bronze-greaved" (7.41), "with bronze cheek-pieces" (12.183, 17.294, 20.397), "bronze-hoofed" (8.41), "inflicted with bronze weaponry" (19.25), and "with bronze tunics" (1.371, 2.47, 5.180, 13.255, 15.330). The great cauldron in which water is boiled (using vast amounts of timber) to wash Patroclus's corpse is made of bronze (18.343–49). Lambasted by Ajax, the Achaeans "fenced their ships round with a hedge of bronze" (15.566–67).[39]

Longer formulae develop the idea of elaborate bronze work: Achilles's chariot is "intricate with bronze" (10.393).[40] And "bronze," used three times over four lines, creates a regular acoustic pattern of guttural consonants to evoke the clanging weapons and groans of combat (13.646–50):

[Harpalion] struck the middle of the shield of Atreus' son with his
     spear
from close, but failed to drive the bronze right through,
and shrank back into his group of comrades, avoiding death,
glancing on every side in case someone should wound his flesh with
     bronze.
But as he drew back, Meriones let fly at him an arrow tipped with
     bronze.

On one occasion a compound with "bronze" even occurs three times in three consecutive lines (12.183–85):

Then the son of Peirithous, strong Polypoetes,
struck Damasus through the helmet with cheek pieces of bronze,
and the bronze helmet did not halt the spear,
but the bronze tip broke through the bone.[41]

Bronze is abundant (17.493), flashing (2.578), long-edged (7.78), shining (12.151), terrible to look upon (13.192), sharp (13.338), stubborn (14.25), pitiless (16.761), and flaming (17.3).[42]

Bronze is also "skin-lacerating" (4.511), a word that makes listeners imagine their flesh being penetrated and torn by an enemy sword. Bronze swords cleave heads and hack off limbs. Bronze spears find gaps between breastplates and helmets and drive painfully through necks and midriffs. Hector does not cease from slaughtering with bronze (17.565–66).[43]

The insistent references to bronze in the poem assault just as power-fully the senses of hearing and sight. The gleaming bronze of the Lapiths' breastplates "clatters" as they defend the Achaean gate (12.151). Loud aural effects are often conveyed by semi-onomatopoeic verbs, such as *bombein* and *brachein*, which begin with the labial consonant "b" and create a similar effect to the English verbs "boom" and "bang." A helmet makes a booming noise as it is dropped to the ground (13.530). When Athena fells Ares, his armor booms all around him (21.408). Bronze "booms terribly" (4.420); arms "boomed with bronze" (12.396); the arms of men "boomed out loud" (16.566). The verb *brachein* reveals an aural affinity between bronze objects, timber, other natural phenomena, and a god's vocaliza-tions, too: it is used when earth "rings" with the din of battle (21.387), a torrent roars (21.9), an oak axle creaks (5.838), and a wounded horse groans (16.468).[44] The verb *brachein* is used for Ares when he attacks Diomedes with his spear of bronze, but Athena diverts the weapon so that Diomedes can retaliate with *his* spear of bronze (5.852–58). Then "brazen Ares," in-jured in the stomach, "bellowed" in pain (5.859) "as loud as nine or ten thousand warriors cry in battle"; "that was how loud was the bellow of Ares, insatiate of war" (5.863).[45] The bronze god's voice booms like the bronze weapons that have afforded him the epithet "bronze."

Menestheus knows that his fellow Achaeans will not be able to hear him as he struggles to defend the wall (12.336–40):

So great was the din, and the shouting went up to heaven,
as shields and helmets with crests of horsehair were struck
    (*ballomenōn*),
and the gates, for they had all been closed.

The sensation of bronze slicing through flesh and bone and the sound of bronze ringing in the ears are equaled by the assault that bronze makes on the eyes. The brilliance of the shining metal in full-frontal engagement is dangerously dazzling (13.339–43):

Battle, which destroys mortals, bristled with long spears,
which they carried for rending flesh. The blaze of bronze

blinded eyes, coming from shiny helmets, and newly burnished
     breastplates,
and gleaming shields as the men advanced in confusion.

The poet continues by remarking that anyone who witnessed this would
be hard-hearted if they took any pleasure in this sight and were not dis-
tressed by it (13.343–44), an emotional prescription that stands out for its
rarity in a poem that consistently seems to extol the "blaze of bronze" and
its lethal consequences.

The verb *lampein*, "gleam," in active and middle forms, recurs in visualizations of arms and combat: intricately wrought arms flash (4.431–32);
a spear-tip gleams (6.319); bronze gleams afar like lightning (10.154); Hector's shield appears and disappears like a star that intermittently shines
through a cloud, and his bronze armor gleams like lightning (11.61–66); he
gleams in his bronze (12.463); spears, helmets, and arms gleam (6.319–20,
16.71, 20.46); even the earth gleams with bronze (20.156). Paris puts on his
armor, elaborately crafted in bronze (6.504), and exults, "shining all over
like the sun" (6.513). Nestor takes up the well-wrought shield "all gleaming
with bronze" that he shares with his son (14.9–11); Achilles is "all gleaming" in his new arms (19.398).[46]

Ajax kills Priam's son Antiphos "of the glittering breastplate" (4.489).
Hector, with "Bronze Ares," kills Oresbios "of the glittering belt" (5.704–7).
Achilles's breastplate, when Patroclus puts it on, is "intricate" and shines
like a star (16.133–34). Hector and Ares both have glittering helmets
(2.816, 18.132, 20.38). Belts, breastplates, and shields can all be "glittering
all over" (4.186, 4.215, 10.77, 11.374, 13.552). Iron, cauldrons, and tripods
can all flash (4.485, 9.123, 24.233). The adjective "shining," *phaeinos*, is applied to bronze (12.151), tin (23.561), a spear (4.496), two different kinds
of shield (3.357, 8.272), Hector's helmet (13.805), a whip (10.500), a belt
(6.219), and the new breastplate that Hephaestus forges for Achilles, which
"shines brighter than the blaze of fire" (18.610). When Achilles finally
grasps his new shield, it generates a massive light like the blazing moon, or
like fire signals from a promontory seen by sailors (19.375–80).[47]

Metaphorical Metal

The heavy, painful feel, glaring appearance, and cacophonous booming
sound of massed bronze armaments on the move and in combat together
constitute one of the most distinctive and memorable aspects of the *Iliad*.
Iliadic heroes are so attached to their armor that they sometimes sleep in

it, their heads resting on their shields (10.150–54). But the existence of metal arms and armor, men looking and sounding as if they were made of metal, and the wielding of metal weapons whether of bronze or iron, also produce in the *Iliad* some strange figurative language.

Apollo urges on the Trojans, reminding them that the Achaeans are not made of stone or iron, but flesh that can be cloven by bronze weapons (4.510–11), with the last line of this formula consisting of four long words of increasing weight that replicate the sensation of physical violence.[48] An "iron din" of battle reaches the bronze heaven at 17.424–25. When Hector is dying, he knows that he cannot persuade Achilles to hand over his corpse to the Trojans, because the heart in the Thessalian hero's breast is as of iron (22.357). When Achilles sacrifices twelve Trojan youths on Patroclus's fire, he sets it alight with "the iron might of fire" to spread at large (23.177).[49] Hecuba accuses the stubborn Priam of having a heart as hard as iron (24.205); when Achilles says the same, it seems more complimentary, as a comment on Priam's courage (24.521).

The narrator even metaphorically wishes he had a "heart of bronze" so as to recount the names of every single soldier (2.490). Hera assumes the likeness of great-hearted Stentor of the bronze voice (5.785), as loud as the voices of fifty other men.[50] Achilles, too, has a bronze voice, which is compared with the sound of the trumpet (18.219–22, an instrument known to Homer's audience but not to his warriors). Iphidamas, his neck slashed by Agamemnon's sword, falls and, in a distinctive phrase, sleeps the profound "sleep of bronze" (11.241).[51] Perhaps what is meant is the sleep appropriate to a man who died encased in or killed by bronze: at 3.57 Hector says that the Trojans must be cowards since they have not long ago put Paris in "a tunic of stone," meaning death by stoning.[52]

A Greek warrior "who donned the artificial human enhancement of bronze chest and leg armour was essentially donning an exoskeleton that replicated the outer appearance of an idealized, 'heroically nude' bronze statue," writes Adrienne Mayor.[53] The Bronze Age experience of wrapping up a body in metal casing, to protect it from metal weapons wielded by foes, deeply affected the human imagination. The Greeks were to invent the myth of Talos, the bronze automaton that Hephaestus made for Zeus to patrol Crete and ward off invaders; he is already illustrated in fifth-century vase painting.[54] The *Iliad* reveals an interest in where the body stops, and the metal casing starts: Agamemnon kills the charioteer Oileus (11.95–98):

He stabbed his forehead with his sharp spear;
his heavy bronze visor could not hold back the weapon.

It penetrated both visor and skull bone,
so that his brains were splattered about inside.

Achilles evokes the image of a corpse being exchanged for its exactly corresponding weight in gold, a motif with precedents in Hittite ritual and Ugaritic epic, when he says even that amount would not be sufficient to ransom Hector's corpse (22.351).[55] There is awareness in the poem that one man's armor might not be a perfect fit for another, even if they are heroes of approximately equivalent physical stature. When Hector puts on Achilles's armor, Zeus makes it fit his body perfectly, and Ares enters into him (17.209–11).

Two-thirds of the way through the poem, we see Hephaestus, the great smith-god, sweating at his anvil, the only proletarian laborer portrayed in any detail in the entire poem, and the sole Olympian artisan. He keeps his fire alight with twenty bellows, his tools neatly lined up in a silver chest, a massive hammer in one hand and tongs in the other. With pre-processed metals he hammers Achilles's great bronze shield, which is elaborately inlaid with gold and silver. And it is when Achilles finally returns in these new arms to lead the Achaeans into battle that extravagant consumption of both timber and metal in the manufacturing of armor is most memorably described. The "brightly gleaming helmets and embossed shields and the strong curving breastplates and the ashen spears" are as dense as snowflakes (19.357–61):

The gleam reached heaven and the flashing bronze
made the whole earth around laugh. And thudding
arose from the feet of men.

In the remarkable series of images that set up the final day of fighting, the vast amount of bronze and weaponry that poured forth from those hundreds of ships is said to make the land around "laugh" (19.362)—a sinister touch, perhaps alluding to the tradition that Earth had begged Zeus to relieve her of the weight of the human race, and implying that nature itself feels vengeful joy in human self-destructiveness. Achilles dons the arms that Hephaestus has made for him: the greaves, fitted with silver ankle pieces, the breastplate, and the bronze sword studded with silver. His great shield gleams like the moon, or a fire on a mountain that sailors can see as storm winds drive them ever farther from land. His helmet with its golden plumes twinkles like a star; he feels free in his movements, the armor like wings that lift him (19.364–86). Man has almost turned into an invulnerable metal warrior automaton, a possibility about which Aeneas envisages Achilles might actually boast.

Aeneas tells the disguised Apollo that Achilles wins incessantly only because the gods assist him: if the gods were to make the terms more fair and even, then Achilles would not prevail over Aeneas easily, "even if he boasted that he was made of solid bronze throughout" (20.102). Soon afterward, Hector develops the theme of Achilles as a man forged out of metal over the flames, saying he will go up against the great Achaean "even if his hands are like fire and his might like gleaming iron" (20.371–72).[56]

Godlike Achilles, for all his grief, feels as if he can fly as he prepares himself for a day of slaughter wrapped in his fabulous metal battle gear. We are reminded of cosmic entities—the moon and stars. Man's relationship with nature is dysfunctional and conflicted. Man at war can resemble cosmic forces; he is also at their mercy. But the secret at the heart of the *Iliad* is that nature is also—for the time being, just for the duration of the Anthropocene—at the mercy of man.

CHAPTER EIGHT

Achilles's Dystopian Shield

ALTHOUGH HUMAN SMITHS ARE occasionally mentioned in the *Iliad*, it is in the unforgettable account of Hephaestus's manufacture of Achilles's new arms and armor that we see the full celebration of metal production in the Bronze and Iron Ages. Hephaestus's status as god of building construction and metallurgy is anticipated by similar deities in Ancient Near Eastern mythology and literature, especially Kothar in the Ugaritic Baal epic; perhaps this heritage is mirrored in the easterly location of Hephaestus's favorite island of Lemnos, near the Asiatic seaboard.[1] There is a fascinating suggestion that the idiom of infinitude in reference to metallurgy was anticipated in Mesopotamian poetry in the account of the products that Kothar forges with his tongs; when the great god Baal asks him to make some gifts for the goddess Athirat,

> He smelted silver, he plated gold;
> he smelted silver into a thousand [pieces],
> he smelted gold into myriads.[2]

But in literature preceding the *Iliad* there is no detailed equivalent of Homer's word-picture depicting Hephaestus in his Olympian residence.

The great Homeric scholar and librarian at Alexandria, Aristarchus of Samothrace (ca. 220–143 BCE), was fascinated by Achilles's shield. He used it repeatedly to illustrate the technology and sociology of the real material world that he assumed underlay Homeric epic.[3] He saw the description of

Hephaestus and his residence, before the commencement of the shield
ekphrasis proper, as dramatic and visually arresting: the poet has "set the
tragic stage" (*proetragōidēsen*). The equivalent verb before the arming of Agamemnon at 11.15–46 is the far more prosaic "prepared" (*proepitēdeuōn*).

"Silver-footed" Thetis visits the technology god in the staggering palace he has built for himself on Olympus. Hephaestus has used his skill to
make a palace for each of the Olympians (1.605–10). But in terms of splen-
dor of residence, he has kept his best work for himself. His own palace is
"imperishable, starry, surpassing those of the other immortals, bronze"
(18.370–71).[4] When Thetis sees Hephaestus, he is sweating, hard at work
among his bellows making twenty tripods, with golden wheels, so that
they can all attend the banquets of the gods of their own accord and return
home again afterward. The tripods are almost completed, and Hephaestus
is putting the finishing touches to them, attaching ornate handles with
rivets (18.373–79).

Hephaestus's wife, who in the *Iliad* is Charis, invites Thetis to sit on a
beautiful, highly wrought, silver-studded chair with a footstool (18.389–90).
Hephaestus recalls his debt to Thetis, incurred when she and the Oceanid
Eurynome rescued him after Hera had cast him out of Olympus to conceal him, ashamed of his lameness (18.395–97). They had taken him to
their submarine cave in the streams of Ocean, which, murmuring with
foam, flowed "infinitely" (*aspetos*) around him (18.403). Hephaestus, who
makes metal objects that far outlast the lives of men, is in close contact
with the boundlessness of the *Iliad*'s imagined universe. He lives in an imperishable palace and was raised in the flow of the infinite Ocean.

The sea goddesses had taught him cunning handiwork, how to make
brooches, spiral armbands, rosettes, and necklaces (18.400–401), so now
he intends to repay Thetis with his hospitality. Homer then offers us a
detailed description, unique in the *Iliad*, of an individual whose day's toil
consists of something other than warfare, changing after work. Hephaestus moves his bellows away from the fire, packs up his tools in a silver
chest, wipes clean his face, hands, neck, and shaggy chest with a sponge,
puts on a tunic, and grabs his walking stick (18.412–16). His maidservants
are golden automata he has made himself; they even have consciousness,
speech, and knowledge of the doings of the gods. They support the limp-
ing god as he moves and sits down on a shining chair (18.417–22). The idea
of the anthropomorphic form entirely made of metal, perhaps suggested
by the advent of bronze body armor, has produced the first robots in ancient Greek literature.[5]

Hephaestus agrees to make new armaments for Achilles, and promptly returns to work (18.468–72):

So saying, he left Thetis there, and went to his bellows.
He turned them towards the fire and ordered them to work.
All twenty of them blew on the smelting-pots,
squeezing out a strong blast in all directions,

We are not told what fuel an immortal smith might use, but the ancient audience will probably have thought of wood and charcoal. The smelting pots will have contained prepared lumps of ore, of kinds that we are about to discover (18.474–77):

He put stubborn bronze and tin on the fire,
and precious gold and silver.
Then he put his great anvil on its stump,
and grasped a huge hammer in one hand and tongs in the other.

Like Lichas, the Spartan searching for the tomb of Orestes in the mid-sixth century BCE, who is described by Herodotus as staring in amazement at what he saw being achieved in the blacksmith's shop into which he had stumbled, we visualize with wonder the greatest smith of them all preparing for mighty labor.[6] We hear no more about the provenance of the primed lumps of ore that Hephaestus has placed in his smelting pots. The audience, however, will have known how much labor—in mines, furnaces, and transport vehicles—had already gone into their preparation.

Plato, according to Diogenes Laërtius, identified mining and logging as the primary *technai* or crafts, taking precedence over carpentry and metalwork, because they produce the materials necessary to the other crafts.[7] Yet most ancient authors, including Homer, erase the experience of mining from the worlds they conjure in their literature. Even basic information about mines in the Bronze Age is in short supply; it is probable, but as yet not certain, that the several deposits of copper ore and lead in Laconia, and alluvial gold in the Eurotas south of Sparta, were already being exploited in the Bronze Age.[8]

Strabo says that in his day it was believed that the wealth of Tantalus and his descendants, the Pelopidae, came from mines in Phrygia and around Mount Sipylus; that of Cadmus originated from mines in Thrace and Mount Pangaion; and Priam's gold came from Astyra near Abdyos on the Asiatic

Enslaved miner at work wearing shackles on an Athenian ceramic. Drawing by Becky Brewis of a painting on an early fifth-century Athenian kylix, now in the Rijksmuseum van Oudheden, Leiden. (Courtesy Becky Brewis)

coast of the Hellespont.[9] Xenophon in *Ways and Means* offers a fair amount of information about the silver mines at Laureion in Attica in classical times, but it is presented entirely from the point of view of those profiting from the industry rather than those laboring in it. It is the much later author Athenaeus of Naucratis (ca. 200 CE) who casually related that the Laureion miners worked in chains and at least once staged a dangerous mutiny; the chains seem confirmed by an Attic kylix of the early fifth century BCE of which the tondo portrays a painfully thin miner, with leg shackles made of metal.[10]

    In an anecdote about a tyrant in the time of Xerxes, there is a casual assumption that miners regularly die while digging for or refining ore; they are sapped of all strength even if they survive.[11] There is only one honest and detailed description of the suffering of mineworkers that has survived from antiquity, and it is Diodorus Siculus's account of the Roman mines in Spain in the first century BCE:

> But to continue with the mines, the slaves who are engaged in the working of them produce for their masters revenues in sums defying belief, but they themselves wear out their bodies both by day and by night in the diggings under the earth, dying in large numbers because of the exceptional hardships they endure. For no respite or pause is granted them in their labours, but compelled beneath blows of the overseers to endure the severity of their plight, they throw away their lives in this wretched manner, although certain of them who can endure it, by virtue of their bodily strength and their persevering souls, suffer such hardships over a long period; indeed death in their eyes is more to be desired than life, because of the magnitude of the hardships they must bear.[12]

When Diodorus was first translated into English in 1700, this passage was used by reformers protesting the suffering of miners in the contemporary English-speaking world.[13]

But there is some serious observation of material reality underlying the complicated narrative scenes that Hephaestus portrays on the shield, and archaeologists tend to locate that reality, in terms of Greek artifacts, in the eighth century BCE. This is when figurative art at both Lefkandi and Knossos, for example, becomes much more prevalent.[14] It is probable that immigrant Near Eastern metalworkers settled in Crete in the ninth century BCE and manufactured, for example, bronze shields decorated with figurative scenes.[15] It is likely, too, that the poet was inspired by looking at particular examples; Hephaestus's technique of highlighting certain figures in silver and gold and the enormous size of his Athena and Ares are both suggestive of conventions regularly found in early figurative art.[16] It is the narrative character of the scenes on the Iliadic shield that distinguish it from other Homeric ekphrases, and these are best associated not with Mycenaean metallurgy but with the relief metalwork on Cypriot and other "Oriental" and "Orientalizing" bronze shields of the Aegean and Crete in the eighth and seventh centuries BCE. These include images of a besieged city and lions attacking a bull, which are found on these actual shields.

In Greek art, the earlier depictions of individual humans, gods, animals,

fishes, and boats are supplemented only in the second half of the eighth
century BCE by complex narrative art in products made in Attica and the
Aegean; vase paintings, and works in metal and clay, begin to show "activity
scenes"—burials, land and sea battles, processions of chariots, and warrior
and funeral games.[17] Phoenician metalwork, however, features embossed
and engraved decorations on imported bowls—including an occasional
scene of a city under siege—found in somewhat earlier, ninth-century
BCE Greek tombs.

There is nothing equivalent to the lawcourt scene on Achilles's shield,
but there are other activities that resemble closely those depicted by Hephaestus. These include ritual activities, which could be weddings, involving women, dancers, and female musicians.[18] There are grazing cattle and
religious processions.[19] There are also military processions, lions, archers,
men playing instruments, and royal figures, occasionally receiving suppliants or tribute.[20] There are hunting scenes, including a depiction of men
being eaten on a hunting expedition, and scenes of feasting.[21] On a silver
bowl from Praeneste, there are both male and female viticulturalists.[22] Many
of the subjects of these images are incorporated by Hephaestus into the
panoramic visions of human life that he so skillfully welds onto Achilles's
new shield. The poets of the *Iliad* may not have been interested in where
metal came from, but as artisans of words they were apparently fascinated
by the visual worlds that an expert Phoenician craftsman could fashion out
of metal alloy.

Achilles's Other and Future Shields

Homer's description of the making of the shield and its ornamental designs, the most famous excursus in the *Iliad*, has so dominated the reception of the epic that the other shield of Achilles, and other versions of the
story, have been overlooked. Homer was already "receiving" and adapting
an older, established epic tradition, and was also partly inspired by scenes
on Phoenician metalwork. He made aesthetic choices. This was not the
first shield that Hephaestus had made for Achilles. When Patroclus dons
Achilles's armor, it includes a great shield with an unspecified emblem
(16.136). The *Iliad* understands this as the shield given to Achilles's father,
Peleus, on the day he married Thetis (17.194–97). But in another tradition,
Achilles himself receives brand new armor when he grows to manhood,
a scene illustrated on several vases, including a two-handled jar (neck-
amphora) now in the Boston Museum of Fine Arts. Thetis and three of

her sisters present the armor to Achilles in the presence of the now mature Peleus: the shield is of Boeotian style and simply depicts the head of a wild-looking Gorgon.[23]

There was almost certainly an on-stage arming scene in Aeschylus's tragic *Iliad*, which contained three plays, the second of which was proba-bly *Nereids*, following *Myrmidons* and preceding *Phrygians*. In the *Iliad*, Thetis is accompanied by her sisters the Nereids when she comes to Achilles after the death of Patroclus, but later she returns alone, bringing him the new armor that Hephaestus has made.[24] Aeschylus merged these two visits into one, making Thetis and the Nereids together bring Achilles his armor. The play seems to lie behind the new series of red-figured vase paintings, inaugurated in the fifth century BCE, showing Thetis bringing Achilles his new armor, but accompanied by that chorus of Nereids, often riding on dolphins, hippocampi, or other sea creatures.[25] Aeschylus seems to have been far more interested in the appearance of this marine chorus than in the appearance of the shield.

We have also lost poetic coverage of what happened to the new shield after the death of Achilles—events that increased its associations with ma-lignity and terror. Ajax, the greatest warrior after Achilles, is disgraced and kills himself because he is not awarded the arms of Achilles, including the shield, due to Odysseus's duplicity and corruption. Neoptolemus, who eventually inherits the arms of Achilles, commits awful, sacrilegious crimes while protected by this terrifying object: Neoptolemus kills Priam, Eury-pylus, Polyxena, Polites, and Astyanax, captures Helenus, and forces An-dromache into concubinage and birthing a new son, Molossus, the ances-tor through Olympias of Alexander the Great.[26]

For in the lost cyclic epic continuing the theme of Troy to the city's fall, Odysseus fetched Neoptolemus from Scyros and presented him with Achilles's armor, apparently because the captured prophet Helenus had said that this was as necessary as the return of Philoctetes from Lemnos if Troy were to fall. In Sophocles's *Philoctetes*, to explain to Philoctetes why he is heading home from Troy in anger, Neoptolemus tells the false tale that he has not been allowed to receive his father's arms from Odysseus.[27]

In his stage version of the *Odyssey*, commissioned by the Royal Shake-speare Company and performed at The Other Place, Stratford-upon-Avon, in 1992, Derek Walcott recalled this detail in *Philoctetes*, and rejected the ancient tradition that Neoptolemus did receive his father's arms. Walcott writes his violent, imperialist, angry Odysseus as if he had indeed refused to concede the shield to Achilles's rightful heir. In the Homeric *Odyssey*, when Odysseus meets Ajax's shade in the Underworld, he explicitly regrets that he won the contest for Achilles's arms, conceding that Ajax's greatness merited them.[28] But Walcott's Odysseus cruelly reminds Ajax that he had beaten Ajax to the shield. The shield in Walcott's interpretation symbolizes Odysseus's difficult personality; he often speaks to it and repeatedly expresses concern about it even when human disasters occur, such as the death of Elpenor. He uses the shield to protect himself, like a turtle's shell, but it is also a dangerous cover that conceals him, emphasizing his status as a liar and a man filled with dark emotions.[29]

Achilles's Iliadic Shield in Allegory and Imperialism

The famous new shield made for Achilles by Hephaestus in the *Iliad*, book 18, was already the subject of diverse interpretations in antiquity and has divided as well as fascinated readers of the epic ever since. It is a symbol of the continuing vitality and contestability of classical epic. Yet it is different in atmosphere from anything else in Homer, as was already sensed by the ancient scholar Zenodotus, who deleted it altogether, deeming it to be spurious.[30] Others in antiquity tried to fix its meaning in specific ways, as representing, for example, the history of Attica: the towns at peace and war supposedly represented Athens and Eleusis, respectively. The customs of marriage and trial by jury began in Athens; the war was the one waged by Eumolpus on Athens. The king at the harvest was imagined to be Triptolemus.[31]

After looking at the history of some other responses to the shield, I propose a new reading fitted to the apocalyptic mood of our era of climate change and widening socioeconomic gaps across the planet. It takes its inspiration from the environmentalist Aldo Leopold's statement that "in dire necessity somebody might write another *Iliad*."[32]

The first designs that Hephaestus puts on the shield are cosmic: the Earth, the heavens, the sea, the unwearied sun, the full moon, and constellations. The second are images of one city's activities: marriages and feasts at which there is dancing in the street and music, but also a noisy scene of litigation in the case of a murdered man. The third is another city threatened by two hostile armies; the citizens are not submitting but are instead planning an ambush. There is an additional party—two herdsmen and their livestock; these are killed by the citizen army, and then a battle breaks out in which supernatural figures of Strife and Tumult and Deadly Fate join in the fighting with the humans.

The next scenes depict arable farming: the plowing of one field, a gang of reapers in another, bushel-binders, and the meal prepared for these laborers. The agricultural section continues with a vineyard scene, followed by herdsmen driving their cattle to pastures. But they are attacked by two lions. This is prior to the mention of a great flock of sheep, whose position in the poem and on the shield makes them seem vulnerable to those lions. The penultimate scene returns to the theme of marriage; Hephaestus designed an image of a dance floor like that which Daedalus made in Knossos for Ariadne. Youths and maidens who are worth many head of cattle dance there. But the last thing the god designed was the "great might of the river Ocean, around the uttermost rim of the strongly wrought shield" (18.478–608). In the most basic sense, the images on the shield are organized in what classical rhetoricians would later label a chiasmus structure: elements in the order ABCBA, defined concisely by Ralf Norrman as "bilateral symmetry about a central axis."[33] On the shield, the chiastic arrangement consists of cosmic elements, humans not at war, humans at war, humans not at war, a cosmic element.[34]

One of the earliest ways to read the shield, which may go back as early as the pre-Socratics, was as a cosmic allegory. The cities at peace and war were associated with Empedocles's primordial principles of Love and Strife; Homer's Hephaestus was compared with the demiurge of Plato's *Timaeus*.[35] But for a detailed allegorization, we need to use the *Homeric Problems* ascribed to the Homeric commentator Heraclitus in the first century CE, and the Byzantine scholar Eustathius in the twelfth. Philip Hardie summarizes the main points: "The god of fire, Hephaestus, is an allegory of the demiurgic fire which creates the universe; the account of the making of the circular shield is an allegory of cosmogony, of the creation of the spherical universe. The four metals of which the Shield is made represent the four elements. Lines 483–5 of the Homeric ecphrasis . . . yield the three world-divisions of earth, sky, and sea, followed by the heavenly bodies. The two cities, one at peace and one at war, are allegories of Empedocles' cosmological principles of *Philia* and *Neikos*. The five layers of which the Shield is constructed represent the five zones into which the earth is divided."[36] Eustathius allegorized the shield strap as the axis that supports the universe.

The allegory probably derives from the Stoic philosopher Crates of Mallos, who worked as librarian and Homeric scholar at Pergamum. He is said to have followed earlier thinkers (that is, Cleanthes) in the view that the Ocean covers the torrid zone between the tropics, but it is certain that he was interested in Homer's Ocean.[37] The Stoic astronomer Geminus said that Crates had fabulously distorted Homer's archaic poetry, and "reads (*metagei*) what Homer said into *sphairopoiïa.*" This last noun, "sphere-construction," is distinctive.[38] It may have meant that Crates designed a terrestrial globe. But it could also have a more abstract meaning—the "spherical structure" of the cosmos or Earth as visualized in the imagination. Crates certainly had a "spherical theory," and this will have had consequences for his exegesis of Homer.[39]

Crates's most famous allegorical readings were of the shields of Agamemnon and Achilles in the *Iliad*, "ecphrastic moments when the narrative stops unfolding and is flooded with spatial coordinates."[40] Crates thought that Homer was attuned to the spherical structure. Eustathius's term *kosmopoiïa* (cosmos-making) "undoubtedly stems from Crates"; Crates was probably the one who first drew a parallel between the two different kinds of making, the shield and the cosmos or sphere; it may have been a response to needing to defend the passage against Zenodotus's deletion.[41] Crates famously called lines 18.481–89 a "representation of the cosmos," "and he gave it a detailed analysis, in terms of zodiacal, climatic, and other heavenly identifications, the common coin of Hellenistic cosmology, and of all cosmology since Plato's pupil Eudoxus in the fourth century."[42] The later grammarian Heraclitus implies that Eudoxus was one of Crates's sources of inspiration.[43] The parallel between the shield and the total universe is a reading to which we shall return.

But the world-making interpretation of the shield was not adopted exclusively by scholars, scientists, and astronomers. It was expropriated for political purposes too. The sun, moon, and stars on the shield may reflect a penchant among archaic warriors to claim celestial, supernatural powers, so as to terrify their enemies. Hardie points out that astral imagery related to warriors recurs in Homer, that Tydeus's shield in Aeschylus's *Seven against Thebes* depicts the blazing heaven, stars, and moon, and that this astronomical comparison was also central to the ruler-ideology of Demetrius Poliorcetes in the third century BCE.[44] For ancient rulers, the cosmos was an irresistible concept in their self-identification as world figures, and they adopted the shield as an emblem of power, the cosmos as *imperium*.

Eustathius supplements his account of Crates's allegorizing of the Shield of Agamemnon as an image of the celestial firmament by saying that this is especially suitable for a king, a *basileus* (on *Iliad* 2.478); in Virgil's *Aeneid* the shield ekphrasis, perhaps drawing on previous uses made of

Achilles's shield by Alexander the Great, becomes a symbol of near-cosmic Roman imperial ambition.[45] This use of shield ekphrasis had a long future in the context of later European imperialism. In the *Columbus*, a 1715 epic by the Italian Jesuit Ubertino da Carrara, there is a long ekphrasis of Columbus's shield and another of his balteus (a wide ornamental belt). The arms are given to Columbus by Aretia, personified Virtue. The shield depicts kings of Spain from 1479 onward, the Spanish Empire in the Iberian Peninsula and Italy; the Indian subcontinent supposedly "discovered" by Vasco da Gama; and, on the concave surface, the New World awaiting Columbus.[46]

And post-Homeric Greek authors of course imitated the *Iliad* shield in their ekphrastic mythical shields, notably in the elder Philostratus's prose paraphrase in his description of a painting of Neoptolemus and Eurypylus in combat at Troy. Philostratus believed that Neoptolemus wielded exactly the same shield as Hephaestus had made for his father Achilles, although, as we have seen, the vase painters were not so sure. Other mythical shields were described by Greek poets, too, from Heracles's shield in the pseudo-Hesiodic *Aspis*, to Quintus of Smyrna's shields of both Achilles and Eurypylus, and Nonnus's shield for Dionysus in his *Dionysiaca*.[47] Latin poets occasionally follow suit, in Statius's shield of Crenaeus in the *Thebaid* and Pallas's shield, described over the course of fifteen lines in Sidonius Apollinaris's *Epithalamium* for Polemius and Araneola.[48] In historical epic, however, the only Latin poet brave enough to follow Virgil was Silius Italicus.

Visual and Verbal Art

By far the most ink has been spent, however, on the Iliadic shield's aesthetic aspects. For the ancients, it became intimately related to Simonides's famous apothegm, "painting is silent poetry and poetry is painting that speaks." These two texts became connected through a common posterity that begins explicitly with the essay attributed to Plutarch, *Essay on the Life and Poetry of Homer*, and passes through the great Renaissance humanist classical scholar Angelo Poliziano (usually known as "Politian" to English-language readers) to Gotthold Lessing, Charles Perrault, André Dacier, and Alexander Pope.[49] In the Renaissance, when Leon Battista Alberti's theory of perspective led to artists believing for the first time that they ought to paint a scene only from one point of view, and at a single moment in time, the shield became central to theories of narrative. For subsequent

The Homeric shield of Achilles reconstructed.
Photograph by Thad Zajdowicz of the Shield of Achilles modeled after
Homer's *Iliad* by John Flaxman, manufactured by Rundell, Bridge,
and Rundell, London, ca. 1821–22. (Courtesy Thad Zajdowicz)

Renaissance thinkers, indeed, the chief interest of the shield lay in their
aesthetic ruminations on the differences between texts and painting. This
interest was expressed in numerous attempts, by authors and artists including Alexander Pope and John Flaxman, to paint, draw, or physically
reconstruct the shield.[50]

The interest continues, via Lessing's *Laocoön*, in the academic preoccupation with ekphrasis, which became so fashionable in the late twentieth century. Thomas Hubbard marvels at how the scene where Achilles's shield is made revolves around the paradox by which literature is able both to describe a work of visual art and to recount action, a paradox that "is in some sense at the root of all great works of art, which constitute themselves as both action and artifact, process and product, becoming and being." For the *Iliad*, "which stands concretized as the first truly 'literary' entity amid a background of ever fluid and changing oral traditions in archaic Greece, this ambiguity and paradox is particularly problematic"; Heffernan writes about how even the most dynamic figures on the shield are related to the static nature of sculptures and to their inorganic condition, heightening what he calls "representational friction, which occurs whenever the dynamic pressure of verbal narrative meets the fixed forms of visual representation and acknowledges it as such."[51] In this passage of the *Iliad*, he writes, Homer becomes temporarily absorbed in the process of "representing representation itself."[52]

Achilles's Shield and the Cold War

Outside classical circles, however, since the mid-twentieth century and especially the beginnings of the Cold War, it has been the sociopolitical contrast between war and peace that has overwhelmingly dominated the reception of Achilles's shield. The man first responsible for this was W. H. Auden, whose poem "The Shield of Achilles" was recognized as a masterpiece from its first publication in 1952. Auden describes Thetis looking over Hephaestus's shoulder, hoping to see well-kept vineyards, olive groves, cities, and ships. But what she sees instead is a wilderness under a leaden sky. There is no grass and no sign of a community, but rather a multitude of people, "a million boots in line." They hear a disembodied voice that uses statistics to prove "that some cause was just," and march away obediently without questioning why.

Instead of lovely scenes of ritual and sacrifice, Thetis sees an atrocity taking place in an enclosure guarded by sentries and surrounded by barbed wire; there are "bored officials" and three pallid figures led out to be bound to posts, watched by an impassive "crowd of ordinary decent" people. The victims have no more control over what Auden calls the "mass and majesty of this world," and they die at the hands of their enemies. Thetis keeps hoping to see scenes of peace—athletics and dancing—but instead there is only a field choked with weeds and a lonely "ragged urchin" who knows of no other world than one in which girls are raped and boys knife one another, where promises are not met and there is no empathy between human beings. Thetis is cast into despair at Hephaestus's handiwork, a gift for her son, "man-slaying Achilles/Who would not live long."[53]

This haunting poem is partly a meditation on the catastrophic effects of war—the million boots, the marching away, the three figures taken to execution, the barbed wire, the ragged urchin, the raped girls and stabbed boys, and the imminent death of iron-hearted, man-slaying Achilles. Some of it reverberates with the despair of Auden, a bohemian originally attracted to communism who had spent a year in Weimar Germany, at the temporary triumph and utter savagery of Nazism; some of it obliquely references his fragile and idiosyncratic Christian Existentialism.[54] The masses persuaded by "a voice without a face" to believe "that some cause was just" without discussing it, and coming to grief, could apply, however, to Nazism, Soviet communism, McCarthyism, or any intolerant doctrinaire establishment religion that Auden, despite his private Christian mysticism, abhorred. The poem responds to the dehumanization of modern society and the petrifaction of the global political world order as the Cold War set in. Deep pessimism emanates from Thetis's failure to locate the scenes of peaceful life that she looks for, and the ragged urchin's inability even to imagine awareness "Of any world where promises were kept,/Or one could weep because another wept."

Half a century later, Seamus Heaney repudiated Auden's infusion of the Homeric shield with dark pessimism. In "A Stove Lid for W. H. Auden," Heaney substitutes, for the military hardware in which Thetis sees reflected only the horror of the Cold War, a metal stove lid that symbolizes all that was most comforting about his private home life. This can protect the poet from his own dark imagination: the world's "mass and majesty" are now encompassed by a simple, "cast-iron stove lid," which Heaney remembers that he, as a boy, could replace safely, keeping control over the destructive fire contained within it.[55] In Auden's poem, the "mass and majesty of this world" have been traduced and turned into hordes of people without even the capacity for independent thought. But Heaney insists on the value of the ordinary, the everyday, and the familiar; this large round metal object, moreover, can stop the very mouth of Hell. Poetry can comfort as well as express despair.[56]

In scholarship, the peace/war antithesis was anticipated by the rather reductive T-scholia on *Iliad* 18.483–606, introducing the shield. They comment that the peaceful city "is intended to make the Trojans consider how many benefits they would have enjoyed, if they were at peace, and the city at war demonstrates their dissension; for in Troy some supported Paris and some Antenor." This analysis shows the commentators struggling with correspondences between the main narrative and the shield, since Troy is not besieged by two armies, and the disagreement between Paris and Agenor is overshadowed in importance by quarrels on the Achaean side. But an uncritical reading of the peaceful city as blissfully untroubled has dominated much more recent interpretations, too. Mark Edwards thinks that the poet repeatedly emphasizes pleasure: "The women stand in their doorways admiring the wedding procession; the lawsuit will be decided by proper legal procedures without further bloodshed, and the fairest judge will win a reward; the basileus watches his workers with joy in his heart, and the ploughmen and reapers receive their refreshment."[57]

This cheering interpretation has been most influentially emphasized by Oliver Taplin, no stranger to Auden's poem, in a much-read article of 1980. This publication date coincided with the height of the Campaign for Nuclear Disarmament of which Taplin was a supporter. He has a vision of the shield's representation of "the good life," the purpose of which is to "make us . . . see [war] in relation to peace," reminding the audience of all that will be lost with the fall of Troy, representing "an easy hedonistic existence spent feasting with the pastimes of conversation, song and dance, making love."[58]

When it comes to our understanding of twentieth-century *history*, however, the relevance of the shield was inscribed on the collective public imagination by the magnum opus of Philip Chase Bobbitt, an American author, academic, lawyer, expert on constitutional theory, and former presidential adviser. *The Shield of Achilles: War, Peace, and the Course of History* had a huge international impact when it was published in 2002. Bobbitt uses the Homeric shield, as well as Auden's reception of it, to make a sweeping historical argument that has become an orthodox way of viewing our present: significant wars have been instrumental in giving shape to states as they evolve from monarchies and territories to nations. But nation-states have used the peace settlements that emerged from wars to change the way that international relations and systems are organized and operate.

Bobbitt says that there was an epochal war lasting from 1914 to 1989 that culminated in the "market state" and the current global order. Within that global order lie the kernels of future conflicts. But these conflicts are less likely to be between so-called legitimate states, which have depended on deterrence to ensure security and peace—instead, weapons of destruction may fall into the hands of non-state agents. Bobbitt's book was published in 2002, before the full impact on the global order of post-9/11 was felt. It was also before most people realized just how imminent our environmental emergency was becoming. It is time, therefore, for a new reading of the shield that addresses our current situation. But first, we need to reach back to the *Iliad* itself to consider the sheer terror that the shield evokes.

The Shield and Human Terror

In an excellent article published in 2003, Stephen Scully warns against being swept away by Taplin's vision of the shield's representation of happy, contented life of peace. Scully also points to the comments about the shield within the *Iliad* and the reactions of its viewers. When Aeneas drives his spear onto it, it is described as "terrible," as "awful to behold" and as containing its own voice, as it groans beneath the spear point (20.259–60). When Thetis brings the arms to Achilles and they clash loudly, fear seizes all the Myrmidons, and none of them dares to look at the arms straight on; they all shrink back in terror. "But Achilles, the more he looked, the more the anger made its way into him. And his eyes, like sun-glare, glittered terribly under his lids" (19.13–17). It is when Hector looks at the bronze in which Achilles is enclosed, which emits rays like burning fire or the sun, that fear seizes him, and he turns to flee (22.134–37).

Scully argues that Achilles's anger when he looks at the shield—an anger that is accompanied by delight—is a response to the shield *in its entirety*. Through its divine framing of human activity within a cosmic perspective, it offers "a distancing vision of the mortal." This is akin to Apollo's terrifying Olympian perspective, when he dismisses mortals as "insignificant" (*deiloi*) and like leaves on a tree that "grow and then fade away" (21.463–66). It is also similar to Zeus's acknowledgment that even though no city on Earth has ever been favored more by him, Troy will fall (4.44–47).

As Scully writes, "The portrait of human existence on the shield and Zeus' image of Troy are statements about the nature of man, not commentary about the good life": Achilles "moves toward a divine synoptic perspective . . . the collective whole transcends human partition. The immortal gods feel no fear at that stark vision" and Achilles is the only mortal who can share their fearlessness.[59] Mortals are terrified by "the sight of the separate bands as part of a unified whole. That synoptic and inhuman perspective breaks the sense of the special status of the human by placing it within the context of a larger cosmos and Zeus' will."[60] Like Victor Turner's description of the participants in "liminal periods of major *rites de passage*," who "are free, under ritual exigency, to contemplate for a while the mysteries that confront all men," Achilles, whom the shield will allow to reassume the identity of warrior, can see the human condition in the objective round.[61] The shield is ultimately an instrument of death "in an impersonal cosmos" that only Achilles, among mortals, can gaze on with pleasure.[62] Gregory Nagy has drawn attention to Telemachus's enraptured response to the bronze gleaming in Nestor's palace in the *Odyssey;* Telemachus assumes that Zeus's house contains such a room, for looking at it fills him with awe.[63]

Nagy suggests that Achilles in the *Iliad* "can make direct contact with the Bronze Age" by projecting a picture from it through the gleam that radiates from its surface.[64] Earlier we saw how Homer evokes not only the feel and sound of bronze, but also the visual impact of bronze armaments, especially those of Achilles when he reenters combat, dazzling, gleaming, glittering in the sunlight of the battlefield, beaming like the moon or stars in the darkness of night. It was doubtless such passages of visual flamboyance that gave rise to the ancient tradition, recorded in the *Life of Homer,* and by the Neoplatonist commentator Hermias in the fifth century CE, that Homer was blinded by the appearance of Achilles at his tomb when the hero appeared in gleaming armor, answering the poet's prayer:

> When he arrived at the tomb of Achilles, he prayed to see the hero
> just as he looked when he entered battle arrayed in his second set of
> armour. When Achilles had appeared to him, Homer was blinded
> by the gleam of the arms.[65]

Poliziano picked up on this tradition in his *Ambra*, a Latin hexameter poem about Homer in the collection of his *Silvae*, which he first read to his students in 1485.[66] Poliziano adapts the ancient tradition, placing the emphasis on the shield. He imagines the poet gazing straight at it:

> That blazing individual showed in the shield the earth and the sea,
> and the unwearying sun, and sun's sister, now in her fullness,
> and stars circling in the silent heaven.
> The poet-priest was transfixed by these, and even while the poor
>   man

incautiously scrutinised each one of them, as he fixed the light of his
    eyes upon them,
night extinguished the light forever.[67]

The moment of greatest physical illumination, and consequently of blinding, coincides precisely with Homer's intellectual apprehension of everything, even celestial bodies, in the universe. Poliziano's Homer now becomes
transformed, and is granted the vatic powers of Tiresias.[68]

My response to the shield resembles that of Poliziano's Homer much
more than the responses of Edwards and Taplin. We are mortals, like the
Myrmidons and Hector: the impersonal depiction of transient human life
is chilling in comparison with those cosmic astronomical bodies and the
ocean encircling the Earth. In 1991, in a more optimistic age, Edwards
could assume that the activities on the shield are paradigmatic representations "of ever-continuing human social activities."[69] But in the twenty-first
century we hear more loudly the contrast between cosmic permanence on
the one hand and the human race interacting with animals on the other;
we recognize the growing denial of people's sense of their special metaphysical status in relation to divinity. We now know we must analyze every
aspect of our relationship with nature and the ways we do agriculture, as
well as our economic and sociopolitical structures, if we are to survive on
this planet.

Many have pointed out the extreme brutality of the war scenes on the
shield, which cast even the poem's descriptions of the Trojan War into a
more rational and aesthetically pleasing light. Hephaestus depicts a mor-
ally murky military situation where a city is besieged by two separate
armies. The most plausible explanation is that one aggressor wants to sack
the city altogether; the other has seen an opportunity to offer to stop the
war in return for half the citizens' possessions.[70] No motive is given for the
warfare other than desire for gain. The citizens refuse both ultimatums and
are preparing to stage an ambush.

As they pour out of the besieged city on the shield, they are led by
Ares and Athena together. But in the *Iliad* these two gods are at irreconcilable odds—Athena objects to Ares's bloodthirsty willingness to jump into
the fray and fight for either side indiscriminately, so enamored is he of
violence. She encourages Diomedes to assault him since he is a berserk
alloprosallos, "one who leaps in to support any side," and a "nasty piece of
work" (5.830).[71] Zeus tells Ares that he is the most hateful to him of all the
Olympian gods, because "strife and wars and battles" always delight him

(5.892). When Athena fells Ares in the battle of the gods, she boasts over his prostrated body that he is a fool who must be mistaken if he believes he is remotely the equal of Athena in might, let alone her superior (21.410–11). Scholars have occasionally noted the difference between the Athena/Ares relationship on the shield and in the rest of the poem, but they have not considered what it might mean.[72] Turning these warring gods into collaborative partners effectively abolishes any distinction between war for a reasoned purpose and mindless savagery.

When full-scale battle is joined outside the besieged city, the involvement of divine figures in slaughtering humans reaches a frenzy of horror (18.535–40):

> In their midst Strife and Tumult joined in, and deadly Fate,
> who was grabbing one freshly wounded man alive, and another
>   unwounded,
> and another she dragged dead through the battle-chaos by his feet;
> and the garment she had about her shoulders was red with the blood
>   of men.
> Just like living humans they joined in the battle,
> and they were hauling away the bodies the others had slain.

Rutherford is so horrified by this scene that he even wants to delete 18.535–38 as an interpolation, although these lines are present in all manuscripts. His main argument is "that they are ill suited to the spirit of the *Iliad* in general and this context in particular"; elsewhere the shield is less extreme and more normal than the narrative of the *Iliad*, he says, but here the description becomes "*more* macabre and horrible."[73] Deadly fate, the death spirit known as the *Kēr*, as he says, must have at least three arms to drag three men simultaneously, and he feels that this monstrous vision does not suit the *Iliad*'s strongly anthropomorphic vision of gods.[74]

Others remove these terrifying figures by claiming that they are a post-Homeric interpolation by someone impressed by their appearance in the pseudo-Hesiodic *Aspis*.[75] Yet Wallace correctly sees that Hephaestus "depicts the pagan deities of annihilation . . . indiscriminately engaged in generating homicidal havoc. The troubling specter of an antinomian anti-community."[76] Indeed, on Scully's reading, this annihilation and dissonance are part of the point. This shield is terrifying because it reveals a cosmic, impersonal view of what is really going on in human conflicts. The illusion that they are rational and can be understood or controlled by merely human powers of deliberation is shattered by the vistas on the shield. What hope is there for humankind if these deadly forces of destruction are uncontrollable?

There are other details on the shield that a twenty-first-century audience might perceive as prescient and sinister. They include the silence of women, who stand at their own doors marveling while others lead brides through the streets and while young men, *kouroi*, dance to pipes and stringed instruments (18.492–96); even the women standing high on the wall of the besieged city make no sound (18.514–15). This muted female presence contrasts with male shouting and loud-voiced heralds in the trial scene (18.502, 507), the herdsmen playing the panpipes (18.525–26), the boy singer (18.569–70), and the noisy lowing cattle and barking dogs, aroused by the youths confronting the lion (18.573–86):

> He inserted a herd of straight-horned cattle;
> the cows were forged from gold and tin,
> and went mooing from the dunghill to pasture,
> alongside the resounding river by the waving reeds.
> Four golden herdsmen paced beside the cattle,
> followed by nine fleet-footed dogs.
> But two terrifying lions, in the middle of the leading cattle,
> clutched a loud-lowing bull. He bellowed loudly
> as they dragged him, pursued by dogs and young men.
> The lions had torn the hide of the large bull,
> and were gorging on the entrails and black blood;
> the herdsmen just kept trying to chase them off, urging on their
>   dogs.
> But the dogs avoided biting on the lions.
> They stood nearby, but out of the way, and barked.

Hubbard enthuses over this scene, which he says shows humans and dogs pulling together to repel the lions, thus displaying "the ultimate triumph of civilization and its values over wild nature, even as the other agricultural scenes show Man harnessing nature to his designs. The development of settled agriculture is nothing less than the fundamental transformative stage in the progress of human culture and is thus archetypal for the preeminence of social values and group work."[77] I do not feel that the text of the *Iliad* I am reading here can be the same one as Hubbard's. The dogs are utterly failing to do what the humans ask of them; the lions continue gorging; the cattle are at the mercy of all three other types of animal. There is an unresolved antagonism between the humans and the lions, and of course there is no sense whatsoever that either species might ever face extinction.

Even the superficially joyous *peaceful* scenes are marred by discordant details that call Taplin's utopian reading into question. For example, six lines of the account of the first town that Hephaestus puts on the shield describe a multiple wedding festivity; but twice the number, twelve lines, describe a much less joyful scene going on simultaneously in the same town (18.497–508). Two men are in dispute about the blood-price of a man who has been killed, and there is a witness, wise man, or judge of some kind they are speaking to or citing, a *histōr*. There is a noisy crowd in the packed marketplace; the people need to be controlled by heralds; there are elders offering their opinions in turn. "And in the middle of them there are two talents of gold, to be delivered to the man who delivered the straightest judgement."

The details of the legal procedure are, moreover, not clear. It is likely that the blood guilt incurred by the killer can be commuted if he makes financial recompense to the family of the victim. Perhaps one man is offering payment while "the other refuses to accept it, or one man claims that he has paid and the other denies that he has received it." They have gone to arbitration; one talent must have been deposited by each of the two parties, and bestowed on the elder whose judgment was endorsed either by the adversaries at law, or by the people as a whole.[78] Either way, this scene has been mightily celebrated as a sign of the emerging civilized values of this town, an early form of independent lawcourt.

The positive interpretation of the litigation scene reached its zenith in its depiction on the seventeen-foot-high bronze (of course) doors of the US Supreme Court, installed in 1935. It is one of the eight panels "that depict the great events in the development of the law and society's achievement of equality under it."[79] The designs were primarily the work of John Donnelly Jr., whose father's firm had received the commission. The Shield of Achilles is the first scene, preceding others including a Praetor's Edict, Julian the Scholar, the Code of Justinian, the Magna Carta, the Westminster Statute, Lord Coke and James I, and Chief Justice Marshall and Justice Story establishing in 1803 the judiciary as a fully independent third branch of the government.

In 1977 the Shield of Achilles panel was described in the *American Bar Association Journal* as a "famous representation of primitive law, signifying

Trial scene from the Shield of Achilles, a panel on the bronze door of the US Supreme Court, Washington, DC. (Photograph by Lucia Kustra, with special thanks to Eric Cline. Courtesy Lucia Kustra)

the original basis of law and custom." The panel depicts "two wise men . . . debating," with "two gold pieces, which will constitute a prize for the man who propounds the best logic and 'straightest judgment,' . . . between them."[80]

Taplin shares the lawyers' enthusiasm for the scene: "we have the stable justice of a civilized city . . . Here is no vendetta or the perilous exile which Homer and his audience associated with a murderer in the age of heroes. We have, rather, arbitrators, speeches on both sides, and considered judgements. The scepter (505) is the symbol of a well-ordered hierarchy."[81] But even Taplin concedes that the scepter "has been somewhat mishandled in the first two books" of the *Iliad*. Worse, the financial details of the dispute are unclear: are the judges the same as the elders? Who will make the final judgment?

The elder Philostratus was baffled by this scene when he tried to reproduce it in his prose ekphrasis *Imagines:* his best guess was that it was an attempt to prevent bribery.[82] Philostratus perceives that the murder trial scene is not a beautiful picture of civilized life, but "represents a state of affairs midway between war and peace in a city that is not at war."[83] I have always found disturbing the implication that legal judgment was financially lucrative. Who is being paid what, and who decides what is the straightest judgment? We could argue that this was a sign of the early corruption of human society, where ethics, justice, and finance were first confused. Wallace calls the trial scene "an archive of bloodshed" and discerns a "disturbing larger pattern . . . a step-by-step deterioration from the promised ecstasy of the marriage processions into homicidal chaos."[84]

The agricultural farming scene (18.550–60) also contains a disturbing detail. Laborers are working hard, reaping and binding sheaves. There is a king in their midst, not working, but taking delight in the produce. Heralds are making a "feast" (*deipnon*) of a great ox they have sacrificed, but the women are preparing the hired laborers' midday "meal" (*daita*) from white barley grain broth (18.559–60). Philostratus tried to ignore or correct this disparity by claiming that the laborers would be fed both meat and barley, but that is not the meaning of the Greek.[85] There is instead extreme social hierarchy; there is ownership of the means of production by a monarch who does not work but eats meat, and there are workers who provide him with his wealth and eat a much more humble meal, as more candid commentators admit.[86] It is important to remember here that in myth the fair division of meat functions to define the archetype of civilized behavior.[87] Achilles calls Agamemnon a dog because he has taken *an unfair portion* of the prizes of war that others have labored to acquire (1.158–60).[88] This is a symbolic representation of the rigid separation of the small elite class, distinguished from peasants and laborers, which, as explained earlier, Ernst Gellner identified as typical of "Agraria," the label he gives to the settled farming societies produced by the Bronze Age.[89]

In the vineyard scene (18.561–72), a boy makes music, but he performs the Linos song. Even scholars who are most upbeat about the scenes of happiness have been struck by this melancholy touch: the Linos song "is always referred to as a dirge, and it seems odd to sing it here on what is obviously a cheerful occasion."[90] The closing tableau of women and men performing a courtship dance (18.590–606), too, which is conventionally diagnosed as offering "a fitting conclusion to the pleasant pictures of human social life which the shield presents," contains a troubling detail.[91] The dancing youths wear golden knives or daggers, *machairai*, hanging from silver baldrics (18.597–98). Aristarchus was disturbed enough to athetize these two lines, since Homer does not elsewhere call a sword a *machaira*, and it is not fitting that dancers should wear a dagger.[92]

The dance is introduced, moreover, by an ominous simile. Hephaestus made a dance floor "like that which Daedalus once made in Knossos for lovely Ariadne" (18.590–92). A famous dance, depicted on the sixth-century François Vase in the Museo Archeologico in Florence, had taken place at Knossos either when Theseus arrived with the fourteen Athenian boys and girls destined to be offered to the Minotaur, or after his successful emergence from the labyrinth.[93]

Arthur Evans, the excavator at Knossos, was obsessed with Homer's reference to Ariadne's floor, because it "positioned ancient Crete exactly where he wanted it to be: as a dream of peace and a magical domain, already legendary by the time of the Trojan War. Evans would return again and again to this scene, locating the original dancing floor at various different sites in and around the palace, reading it into dozens of different artifacts, and developing an ever more elaborate reconstruction of the dance itself."[94] On a common feast day, he insisted that his workmen, both Christian and "Mahometan," dance this supposititious Cretan *choros* "in the Western Court of the Palace."[95] But the implicit reverberations of the violent death of marriageable young people, sacrificed to a terrifying monster, and the ill-starred liaison of Theseus and Ariadne, inevitably render this an uncomfortably sinister simile to place in our heads before we hear the description of the dance.[96] It also, retrospectively, throws dark light on the marriage scene near the beginning of the ekphrasis.[97]

The Round River

The earliest readings of Achilles's shield were therefore cosmic allegories. These were followed by political applications, afterward by a long-lasting aesthetic preoccupation with what the shield had to say about the differences between visual and verbal art; and later, in the eighteenth and nineteenth centuries, by archaeological reconstructions. The aesthetic preoccupation never disappeared, although it was challenged in the second half of the twentieth century, after two world wars, with a return to a sociopolitical interpretations. The twentieth and early twenty-first-century readings of the shield were dominated by Cold War ideology and fears of boots-on-the-ground international military combat. But over the past few years this anxiety has been replaced by our fears that our industrial, capitalist, patriarchal way of life is unsustainable, and that *Homo sapiens* may be on the brink of extinction. By noticing on Achilles's shield the dark undertow even in the cities at peace, by adopting its synoptic view of humanity's insignificant place within the cosmos, and by resuscitating Crates of Mallos's approach to Achilles's shield as a cosmic allegory, can we repurpose it as something good to think with in the new age of apocalyptic terror?

There is just one example of an apparently happy, peaceful community whose mode of production is sustainable on an ancient shield ekphrasis. It was the shield made as a gift for a historical figure in Latin epic, Hannibal in Silius Italicus's *Punica*, by the people who live by the Atlantic Ocean.[98] This poet described in some sixty-two lines (against the 103 lines of Virgil's ekphrasis in the *Aeneid*) the shield that "the peoples who dwell by the Atlantic" had brought as a gift for Hannibal.[99] Its scenes of pre-urban peace contain a utopian picture of a hunter-gatherer, and a nomadic mode of production before built cities, before social hierarchies and the privatization of land, before cattle farming, and before the mass extinction of wild animals. It contains few of the off-key notes that resound in such a sinister way from Achilles's Iliadic shield: a North African woman, admittedly presented in a slightly sinister way, soothes lionesses in her native tongue, and the Punic shepherd's flocks roam freely over the plains, their limitless pasturelands being forbidden nowhere and to nobody; this nomad carries all he possesses with him—his javelin, his dog, his tent, his tools for sparking fire from flint, and his reed pipe.[100] This is another ancient shield, ultimately inspired by Homer, with which we could marginally better arm ourselves to face our battle for survival.

But more conceptually powerful is an idea suggested by the environmental activist, forester, and philosopher Aldo Leopold, the son of a German immigrant into Iowa; he became a professor at the University of Wisconsin in the early twentieth century. Leopold was influential in the development of modern environmental ethics and in the movement for wilderness conservation. He saw ecology as the fusion point of science and land community. He developed a conservation ethic, a "land ethic."[101]

Leopold's ethics had a profound impact on the environmental movement. In his popular, accessible writing, he used a range of imagery, attempting to imagine an alternative way of inhabiting the land. He argued that we need to "think like a mountain," to acknowledge that killing even one animal has serious implications for the ecosystem. Leopold's writing is replete with images of the land, and he uses numerous different metaphors to describe the natural world as an organism, a community, a balance, a pyramid, an engine, and a song; farms are presented as the "stages" trodden by farmers among other "players" (such as animals), with each living out their various "dramas."[102] Leopold's most famous quotation is "A thing is right when it tends to preserve the integrity, stability, and beauty of the biotic community. It is wrong when it tends otherwise."[103] In the public imagination, his theories have been overshadowed by the Gaia hypothesis, developed by James Lovelock and Lynn Margulis in the 1970s, which drew on classical mythology to propose that living organisms interact with their inorganic surroundings on Earth to form a synergistic and self-regulating, complex system that maintains the conditions for life on the planet.[104] But Leopold's image of the Round River has equally classical credentials and can help us use Homer's shield, with its images bound by the circle of Ocean, in an ecologically suggestive way.

In "The Round River: A Parable," Leopold used a story about a celebrated figure in American folklore, Paul Bunyan, whose orally transmitted adventures were committed to writing by James MacGillivray.[105] In the most famous of these stories, "The Round River Drive," first published in the *Detroit News Tribune* in 1906, Bunyan embarks from his camp on the riverside to drive logs down an unknown Wisconsin river.[106] After some weeks, he realizes that he has just passed the same camp from which he first departed, so the river must in fact be round. Leopold turns this round river into a parable for the never-failing ecological process: "The current is the stream of energy which flows out of the soil into plants, thence into animals, thence back into the soil in a never-ending circuit of life."[107]

Leopold thinks that this image of the Earth as a circular stream could help transcend the disciplinary separation of biology (flora and fauna), geology (how the land was formed), and agriculture and engineering (how to exploit them). The Round River can help us "convert our collective knowledge of biotic materials into a collective wisdom of biotic navigation."[108] An invention of American folklore, the Round River is a non-hierarchical gyratory loop in which the life and death of even the top predators are but moments in the basic circular flow. Our problem, as *homines* who call ourselves *sapientes*, writes Leopold sardonically, is that we have interrupted that flow for our own advantage: "We of the genus Homo ride the logs that float down the Round River, and by a little judicious 'burling' we have learned to guide their direction and speed. This feat entitles us to the specific appellation sapiens."[109]

Leopold was steeped in Homer, and drew on his epics. His "The Land Ethic" begins, "When god-like Odysseus returned from the wars in Troy."[110] In *A Sand County Almanac*, a section entitled "Odyssey" describes the circulation of atoms around an ecosystem as a journey comparable to Odysseus's departure from and return to Ithaca.[111] The environmental philosopher Henry Dicks has also pointed out the resonances connecting Homer's Ocean and Leopold's Round River. Ocean was the primordial deity, a self-maintaining, circular flow. As the Achaeans and Trojans prepare for battle, Zeus summons to Mount Olympus all the gods but one, Ocean, because he holds everything together.[112] The text immediately following Leopold's "The Round River" is "Goose Music," in which Leopold writes that "in dire necessity somebody might write another *Iliad*."[113] As Dicks concludes, "Leopold may not have written another *Iliad*, but there is a strong case for saying that out of the 'dire necessity' of preserving the land from destruction he forged another Shield of Achilles. On this view, Leopold is not so much a contemporary Odysseus as a contemporary Homer, a man who took perhaps the most famous folkloric story of his native country and, from that story, brought forth a world."[114] The scenes on the shield may depict human activities in all their profligate senselessness, but if we learn to respect the round river whose flow encircles them, there may yet be a future for life on earth, after all.

CHAPTER NINE

Humans, Rivers, Floods, and Fire

Divine Scamander, purpled yet with blood
Of Greekes and Trojans, which therein did die

—EDMUND SPENSER, *Faerie Queene*, book 4, canto 11

AT THE SUPERNATURAL CLIMAX of the *Iliad*, the River Scamander, whom the gods call Xanthos, asks Achilles to stop slaying Trojans in his streams. His watercourse is now blocked, and he cannot get past the corpses to flow into the sea. The Ionian Greeks knew about blocked rivers. The coastal forests were destroyed by the archaic age; the loggers had moved inland up the three major river basins of southwest Anatolia—Maeander, Caicus, and Cayster—and turned them into plowland. The earth had begun to crumble and erode, and the Maeander was silting up at an alarming rate, eventually leaving former port towns landlocked.[1] But Achilles is not to be halted, and the River, at last, after more than two hundred lines depicting slaughter in his shoals and eddies, brutally retaliates. He rouses all his waters and rushes against Achilles, his floodtide now sweeping away the massed cadavers (21.214–39).

Achilles has finally met his match. He has reached the limits of his physical strength. Man can maltreat rivers for prolonged periods, but there will, inevitably, be a price to pay. Achilles cannot withstand the enormous wave and becomes unsteady on his feet. So he grabs an elm tree, but in doing so uproots it. The tree falls across the River, bringing with it the soil of the bank in which it had been planted. Achilles leaps from the River and tries to escape across the plain, but he is filled with terror. The River pursues him with a mighty roar (21.240–56).

Here Homer illustrates Scamander's violence by introducing the crucial simile, quoted previously, where a gardener tries to divert a stream of water, but in response it runs out of control (21.258–64). Anthropogenic interventions and global warming over the past three decades have already inflicted damage on the streamflow and precipitation levels of the Scamander, while evaporation and the river's temperature are increasing; given the phosphates leaking into the river from farms along its banks, environmentalists are also pleading for management plans to ensure that nutrient concentrations and phytoplankton biomass are monitored through seasonal sampling of sites along the river's course.[2]

In the *Iliad*, the River continues his pursuit, and Achilles is incapable of standing his ground against the ineluctable waves that beat from above on his shoulders and drag away the soil from beneath his feet. Achilles now appeals to Zeus. He had been promised a glorious death in combat (21.281–83),

> but now it is fated for me to die miserably,
> cut off like a boy swineherd trapped in a great river,
> when a torrent sweeps him away as he tries to cross it in winter.

If Scamander is like a stream responding with unexpected force when a man tries to control it, Achilles is like a boy utterly helpless in the face of a river. These two similes—framing the sole episode in the *Iliad* where Achilles the godlike superman, fired by a godlike fury, *mēnis*, is about to be vanquished—are produced by the poetic imagination of an age when men believed that there were no limits to the provision of resources guaranteed to humans by nature.

Moreover, book 21 of the *Iliad*, the poem representing the apogee of Achilles's wrath and its frustration, offers us an extraordinary sequence of poetry and aesthetic experience. This was noticed in earlier centuries, leading to several important attempts to depict it visually by significant artists, and its specification as the theme for both the Propyläen Prize and the Prix de Rome in 1831.[3] It is therefore astonishing, as Holmes has pointed out in her excellent study, how little attention this episode has attracted in scholarship recently.[4] It is as if Homer had thrown, into a kaleidoscope, diverse colored fragments of glass, some representing humans interfering with the rest of the natural world, others representing humans being destroyed by natural disaster—the elements of fire (both wild and domestic), water (both rivers and sea), and wind. In addition, there are wild glass fragments representing gods who play with both humans and elements as they jostle for authority between themselves. We witness cataclysmic flood and conflagration, the wholesale destruction of life—botanical and zoological as well as human—in scenes unparalleled in the rest of the poem; they offer a vision of what the world might become if Hera does not harness wind and fire, personified as Hephaestus, Zephyros, and Notos (21.331–41), to put a stop to this elemental aquatic apocalypse, merely in order to save Achilles to fight another day.

The episode describes what is happening right now in this single river valley, but also leaves stamped on our consciousness unforgettable visions of the annihilation of humanity by river, wind, and fire. And various literary means comingle the individual elements and agents to make it almost impossible to individuate them: Achilles burns and blazes like fire (21.12), but the river has silvery eddies (21.8). This in turn obscures any clear understanding of cause and effect—the full extent of human responsibility or helplessness, and the full extent of the autonomy of natural forces; they are able not only to fight back against human interference, but also to act spontaneously.

The elemental showdown provides the climax to Achilles's day of utter rage after he learns of the death of Patroclus and is engulfed by a black cloud of grief (18.22). He collapses in the dust, smearing his face with ashes and tearing his hair with both hands (18.23–27). His lamentations and those of the slave women seized by himself and Patroclus arouse his mother, Thetis, who emerges from the sea to approach him; he tells her that life will mean nothing to him until he has killed Hector. Thetis responds, weeping, that his own death shall follow swiftly on Hector's (18.95–96). But Achilles tells her not to attempt to dissuade him from avenging Patroclus, prompting her to go to Hephaestus to ask him to make new armor for her son (18.145–47).

Their extraordinary dialogue occurs the day before Achilles's incomparable, terrifyingly berserk rampage, thus setting it up as one that will lead as inevitably to his own death as to Hector's. This casts a bleak metaphysical light on all that is to ensue. The hero refuses his last chance to save himself from death at Troy, and elects to die after winning glory.

While they have been speaking, the battle over Patroclus's corpse has continued unabated on the battlefield, and Hector is about to gain control of the body. Hera sends Iris, the rainbow goddess, to urge Achilles to action, even though he has as yet no new armor. Iris tells him that he can nevertheless go and show himself to the Trojans from a distance, to implant terror in them and perhaps grant the Achaeans some breathing space (18.165–80).

Athena protects the unarmed superhero with her aegis, a golden cloud over his head and a blindingly bright fire issuing from his body. Man and elemental force become indivisible (18.203–6). With divine aid, Achilles virtually turns into the element of fire itself. Here a long simile describes the visual effect of his supernaturally enhanced appearance (18.207–13):

> When smoke emanates from a city and reaches far to the sky,
> from an island which enemies are besieging,
> and they contend all day long in hateful war
> from their city walls, at sunset
> the beacon-fires flame up one after another, and the glare
> springs up high for people who live nearby to see,
> in case they can come in their ships to ward off harm.

Achilles looks exactly like what he represents: the living instrument that will ensure that the besieged city of Troy will be placed in parlous danger of incineration.

Looking as if he were physically aflame, he strides to the trench and bellows; Athena joins in his shouting. Three times he bellows, stupefying the Trojans and their horses. They are thrown into confusion by the volume of Achilles's "bronze" voice, which sounds "as clear as a trumpet when murderous enemies are besieging a city" (18.219–20). In the chaos, twelve of their warriors die at the hands of their own fellow soldiers, from the equivalent of "friendly fire." At last the body of Patroclus is retrieved by the Greeks, and Hera makes the sun go down (18.239–42). The Trojans bivouac for the night outside the city, while the Achaeans prepare Patroclus for his funeral. Meanwhile, Hephaestus forges the new arms for Achilles, including the shield.

The next day, Achilles and Agamemnon resolve their differences, but Achilles swears that he will touch neither food nor drink until he has avenged Patroclus. Zeus is concerned and sends Athena to bestow nectar and ambrosia upon him, to keep him from hunger (19.347–48). Achilles has already been turned visually into the equivalent of fire. Now his separation from normal human experience, his temporary participation in the realm of the divine, is further emphasized by his unforgettable arming scene (19.365–74):

> His teeth could be heard gnashing; both his eyes
> blazed like flaming fire, and unbearable grief
> sank into his heart. In his rage against the Trojans
> he donned the gifts of the god that Hephaestus had laboured to
>   forge for him.
> First, he put round his shins the fine greaves, fitted with silver over
>   the ankles.
> Second, he fixed the breastplate round his chest,
> and cast the bronze sword, studded with silver,
> over his shoulders. Then he took hold of his great solid shield,
> and from it emanated a gleam like the moon.

In a second, longer, and far stranger simile, the gleam coming from the shield is likened to a natural element that has been tamed by nature—a fire burning in a human residence—but which causes emotional pain to other humans who cannot tame nature (19.375–78):

> It was as when the shining light of a burning fire can be seen
> from the sea by sailors, and it burns high up in the mountains,
> in an isolated refuge; but the storm-winds carry them against their
>   will
> out across the fishy sea, far away from their friends.

Achilles's war equipment may be beautiful, but it bodes misery. The fire that men have made to help them survive has a far more sinister additional function: it illuminates all that the men who have tried to tame the seas in their ships can lose if natural forces are not compliant. Man's cultural inventions will one day illuminate, when he is in jeopardy, the damage he has done to nature. Achilles completes his arming scene by donning his helmet that shines like a star and grasping his spear of Pelian ash—foreshadowing the tragic deforestation, mainly by wildfires and consequent erosion, of the soil of Pelion in Thessaly.[5]

The next book opens with Zeus asking Thetis to convene a divine assembly. A strange emphasis is placed on the invitation being extended not just to the gods but also specifically to all the water gods: we are told (20.7–9),

> And so not one of the rivers was absent except Ocean,
> nor any of the nymphs who inhabit fair sacred groves,
> and river-sources and grassy meadows.

Ocean was exempt. A self-maintaining, circular flow, Ocean was the last thing that Hephaestus had added to Achilles's shield, to encircle and contain with it all the other cosmic and human scenes he had painstakingly forged and crafted (18.607–8).

Ocean is sometimes seen as the eldest of the Titans, but other archaic traditions grant him a primordial role in the creation of the universe, making him far more ancient than Zeus and the other Olympians. For example, when Hera asks for Aphrodite's assistance in her planned seduction of Zeus in book 14, she invents a reason why she needs erotic help. She says she is going to visit her currently estranged foster parents Ocean, the "origin (*genesin*) of gods" (14.201, see also 14.302). Hypnos calls Ocean "origin (*genesis*) of everything" or "origin of all streams" (14.246). Aristophanes records a tradition that the three primordial beings, brought into existence by Eros, were Ouranos, Ocean, and Gaia.[6] The Orphic *Hymn to Ocean* calls him "imperishable and eternal" and the "origin of both gods and men."[7] In Greek literature, Ocean's status relative to Zeus's must always be ambiguous.

In this council of all the gods but Ocean, Zeus says that the gods may return to the field of battle and assist the side each favors, although they soon take their seats to watch. As the battle begins, Achilles searches for Hector. He nearly kills Aeneas, but Poseidon, despite supporting the Achaeans, rescues Aphrodite's son (20.318–29).

Achilles kills several Trojans. When he kills Hector's brother, Polydorus, Hector charges against Achilles and needs to be rescued by Apollo. Achilles continues slaughtering Trojans in the most savage *aristeia* in the entire *Iliad*; some of his victims are associated with rivers or divinities of the elements—Iphition, the son of a Naiad and Otrynteus, born in the lands of his ancestors by the Gygaean Lake, the River Hyllus that teems with fish, and eddying Hermus (20.381–92). The slaughterhouse narrative of book 20 concludes with two similes helping us understand the sheer force of the furious energy driving on Achilles. He is sequentially compared *both* to a force of nature destroying all in its path without the involvement of humans, *and* to a man harnessing nature to compel both animals and plants to render him a livelihood. He charges on (20.490–92),

> As a god-kindled fire rages through the deep ravines
> of a parched mountainside, and the deep forest burns,
> and the whirling wind drives the flame on in every direction.

The horses drawing his chariot trample on corpses and shields, spattering blood everywhere, like the bulls a man yokes "to tread white barley on a well-built threshing floor, and the husks are quickly peeled beneath the hooves of the bulls as they bellow loudly" (20.495–97). Natural disaster? Or man forcing a living from nature? In his moment of abject fury, Achilles seems inseparable from either.

Achilles vs. Scamander

With these similes, Achilles rages on into the next book, driving the Trojans forward, and arrives in its first line at the fair-flowing river Scamander/Xanthos, who, we are reminded, is the son of Zeus himself (21.2). Half of the Trojans are forced to jump into the river, like locusts driven on by a wildfire (21.12–16). Achilles leaves his spear by a tamarisk bush and leaps in after them with only his sword. Their hideous groans can be heard as the river runs with their blood. They are like fish trying to escape an enormous dolphin (21.17–26).

Achilles's rampage pauses as he captures the twelve young Trojans he intends to sacrifice on Hector's pyre and sends them to the ships. But then he resumes his killing spree. Details we are given concerning the last two warriors to be slain keep our focus on the twin notions of natural catastrophe and human exploitation of nature. One is Lycaon, a son of Priam whom Achilles had previously captured when the youth was cutting shoots from a wild fig tree (a nonrenewable resource) in his father's vineyard. He was going to use them to make the wicker sides of a chariot (21.34–39). Achilles had sold Lycaon into slavery, but Lycaon's release had been secured by a ransom of over three hundred cattle. Achilles kills him and hurls his body into the river, saying that the fish will lick the blood from his wound, and Scamander will carry his corpse down to the open sea where more fish will eat Lycaon's white fat (21.125–27). As Holmes memorably puts it, the River "supports life at a level where the boundaries between person and animal are collapsed. In the economy of consumption at work in its waters, flesh is flesh: life carries on indifferent to the corpse, fish eating the human beings who eat fish."[8]

We are explicitly told that Achilles's treatment of Lycaon's corpse makes the River even angrier (21.136–38). But the River does not launch himself into open conflict with Achilles just yet. A recent arrival at Troy is Asteropaeus from Paeonia in northern Macedonia. He is the grandson of the northern Macedonian and Greek River Axius (21.140–43; this river is now called the Vardar and is heavily polluted with pesticides).[9]

Asteropaeus stands forth from the river to face Achilles, and the river god puts courage in this grandson-of-a-river's heart (21.145–46). In the warriors' pre-combat altercation, Asteropaeus stresses his descent from a mighty river (21.154–60). He does not shame himself in combat, either. He is ambidextrous and is unique in the *Iliad* in that he succeeds in drawing blood from Achilles when he hits his forearm (21.161–68), though Achilles prevails of course. In his boast over the corpse, Achilles reintroduces the focus on his victim's descent from a river. That may have made Asteropaeus mighty, but Zeus is mightier than rivers, so descendants of Zeus must be mightier than descendants from rivers. Achilles includes Xanthos in his taunt, suggesting that Asteropaeus should ask him for help, but that it might not be a good idea (21.184–93).

Achilles now makes a fascinating claim about relative status in the universe of rivers, Ocean, and the Olympian Zeus. Achilles asserts that neither Achelous, nor even Ocean, claims equality with Zeus (21.193–95, *isopharizei*). Even Ocean is afraid of Zeus's lightning and thunder (21.198–99). But this is a rhetorical claim, not a statement of fact. The ease with which the river—who is himself a son of Zeus—is about to vanquish the great-grandson of Zeus and son of Thetis certainly calls into question the veracity of Achilles's claim.

The fish devour the fat around Asteropaeus's kidneys, and Achilles dispatches seven more Paeonians (21.203–10) before Scamander finally reaches his breaking point and addresses Achilles directly. The deep-rippling River "likened himself to a man, and spoke from out of the deep ripples" (21.213).[10] Holmes argues that Scamander is not to be seen as physically embodied, that "the voice rises from the eddies, indicating that Scamander's anthropomorphism is limited to his speech," but I cannot agree.[11] All the rivers except Ocean attended the divine council at the opening of the previous book; they need to be imagined anthropomorphically to make sense of Homer's statement that they arrived at Zeus's house and sat down in the polished colonnade that Hephaestus, with his cunning skill, had crafted

(20.10–12).[12] It is interesting that Scamander bellows like a bull (21.237), which could be an acknowledgment of the convention of representing rivers, especially those in combat, as bulls or with bulls' heads.[13] It may also suggest how much the entire sequence owes to Mesopotamian deluge-and-flood narratives, in one of which the flood is said to be noisy like a bull.[14] But Achilles certainly "finds himself engulfed in a realm of watery liminality. The surrealistic description of the river's power underscores its ambiguous divinity, that is, its power both to destroy and save."[15]

It is at this point in the action that we rejoin the narrative of the *Iliad* where this chapter began: Achilles cannot equal the river in combat, and two similes point to man's inability to control nature without unforeseen consequences, and man's utter helplessness in the face of a river that without human intervention has become a total threat to human survival.

We have reached not an impasse, but the moment when the mighty Achilles seems inevitably to be about to be destroyed by the river god whose course he has obstructed and who has now broken his banks to pursue Achilles across the plain (21.233–39). Human excellence and human technology in the forging of arms have met their limits in the form of retaliation from the natural world. Only supernatural intervention can prevent Achilles from succumbing now, and it comes in the form of Poseidon and Athena. They assume human shapes and clasp his hands. Poseidon tells him that the River will soon give him respite, and that he must continue to drive the Trojans back into Troy (21.284–97).

Given new strength by Athena, Achilles now faces the River again, and the River has difficulty withstanding him. But the River has not run out of resources. There are other rivers to call upon. Scamander is still full of rage (*menos*) and in his wrath raises himself on high, turning his surge into the form of a crest; he calls on his brother River Simois to his aid.

In one of the most terrifying speeches in the *Iliad*, river god exhorts river god (21.311–14):

> Come quickly to help me. Fill your streams
> with water from the springs, arouse all your torrents,
> raise a huge wave, stir up a mighty roar
> of tree-trunks and stones.

Scamander predicts that Achilles's strength, beauty, and even his extraordinary weaponry will not be able to save him (21.317–21):

The river gods confront Achilles. Drawing by Becky Brewis of an
anonymous engraving ca. 1752 and now in the Louvre, depicting Achilles
with the river gods Scamander and Simois. (Courtesy Becky Brewis)

> They will lie deep beneath the marsh
> covered up with slime. And as for him, I will enfold him
> in sands, pouring over him a layer of silt
> beyond measurement (*murion*). The Achaeans will not know where
> to gather his bones, under such a depth of mud I will conceal him.

This is one of the several images the *Iliad* offers us of all traces of human
activity being erased and hidden from later view by the action of elemental
forces and gods—the destruction of the Achaeans' wall by Apollo and the
obliteration of the Greek fortified camp in later times by Apollo and Poseidon. These images implicitly negate the poem's claim to lend immortality to the heroic warfare it narrates by visualizing for the listener a physical world where every single sign that humans had ever existed has been
wiped off the face of the Earth.

Hephaestus vs. Scamander

After summoning Simois, Scamander gathers all his force and is about to overwhelm Achilles (21.324–27):

> He spoke, and rose churning and seething from on high against
>   Achilles,
> boiling with foam and blood and corpses.
> The purple flood of the heaven-fed river stood
> towering above him, and was about to take down the son of Peleus.

It takes another Olympian to rescue Achilles from the River's redoubled fury. Hera now intervenes. She tells her son Hephaestus to send forth fire, which she will exacerbate by rousing western and southern winds from the sea to fan the flames. The fire will destroy all the Trojan dead and Trojan battle gear; Hephaestus is to burn up the trees that line the River's banks and not cease until she tells him to (21.328–41).

Hephaestus obeys. The fire soars across the plain, burning all the corpses and drying out the soil. In a peculiar simile, the process is likened to the effect of the North Wind, Boreas, drying out a rain-sodden field at harvest time, randomly bringing joy to its farmer: natural events for which men are not responsible can work in their favor as well as to their detriment (21.346–49). But now Hephaestus turns his flames against the River. Here the poet produces the very opposite of the familiar classical *locus amoenus* describing idyllic pastoral scenes, as he lends serious pathos to the destruction of flora and fauna by fire (21.350–55):

> Burning were the elms and the willows and the tamarisks.
> Burning were the lotus and the rushes and the galingale,
> that grew in abundance by the River's lovely streams.
> In sore affliction were the eels and the fishes in the eddies;
> they were tumbling in all directions, tormented by the blasts of wily
>   Hephaestus.
> Burning, too, was the mighty river.

This is a truly dystopian vision of organic life being annihilated by a fire. It is, ultimately, nature's response to gross pollution caused by human acts of warfare.

The climactic half-line that says that even the mighty River himself

was alight (21.355) must today prompt memories of the appalling inci-
dences of burning rivers caused in recent times by human pollution, usually oil or chemical spillage. The Meiyu River in eastern China burst into
flame on March 5, 2014. Its waters were fearfully polluted by chemicals,
oil, and untreated sewage from factories upstream. The flames reached
over sixteen feet high and destroyed cars parked along the riverbanks.[16]
Earlier, in 1969, the Rouge River in Detroit, which is encircled by petro-
leum industry works, caused nearby oil tank storage buildings to explode
and rendered the city's air unbreathable.[17] There are many more shocking
examples.

But at this point in the *Iliad*, Scamander gives in. He realizes that it is
not worth sacrificing his own interests for the good of saving the Trojans.
What is it to him as a River, responsible only for his own ecosystem, if some
humans defeat some others? The gods now go to war with each other until
it is time for Achilles finally to face Hector. But the real point of the Scamander sequence is that the battle between Achilles and the River, and the
River and the Fire god, aestheticize the eventual consequences of the entire
wasteful mode of production—Hephaestus was last seen bronze-smelting
for war but now plays the key role in destroying the whole natural environment and wrecking the entire Scamander ecosystem. None of this would
have been necessary had the Achaeans never made war on Troy in the first
place. None of this would have happened if Achilles had not extended the
effects of his rage beyond humanity to the natural world and so carelessly
polluted and distorted the very watercourse of the River Scamander.

By developing an ecological consciousness, and interpreting the furor
of *Iliad* 21 ecocritically to expose its environmental unconscious, we can
perhaps read Achilles's wrath in a way that can help us begin to heal the
rifts between humans and the environment, and develop a better way of life
in which Homer's surreal, phantasmagoric, and dystopian visions of elemental strife and chaos never become actuality. In this better way of life,
humans, animals, plants, rivers, and fire can flourish in reciprocal cooper-
ation rather than antagonistic mutual disruption. There must be a way to
relinquish our rage, and make watercourses through our gardens that do not
do such damage to our rivers that they retaliate like Homer's Scamander.

Epilogue

The Greeks of the archaic age had already asked whether the Trojan War had been designed by Zeus to reduce the human population in order to relieve Mother Earth of her burden. The *Iliad*'s apocalyptic visions of elemental cataclysm and the erasure of human civilization are responses to a real sense of precarity and fear of natural catastrophe in the future. Reading the poem in detail reveals anxieties about seismic events, tsunamis, storms, floods, wildfires, plagues, and famines that were structural to the archaic Greek imagination, as they were to other Ancient Near Eastern texts, and are once again to our twenty-first-century world. But the poets of the *Iliad* may have been responding more specifically to their awareness, drawn from poetry and storytelling as well as material remains, of the great Mycenaean palace civilizations that had collapsed in Greece and Crete at the end of the Bronze Age. The *Iliad* contains distant memories of everything the Mycenaeans had suffered before their civilization disintegrated—famine, plague, fire, flood, menacing waves, earthquake, whirlwinds, and destruction of the works of man. Its poets' visions of apocalypse therefore look forward as expressions of anxieties about potential future catastrophe, as well as backward to remembered reality.

Yet also central to the ideology of the *Iliad* is the idiom of infinitude, an assumption that the physical Earth, its contents, and the resources needed by humans, are somehow limitless. The implied infinity is temporal as well as spatial and quantitative. The glory of the heroes, who were larger and stronger than those of today, will be forever unperishing. The *Iliad* knows no possibility of the extinction of the entire human race. Most ancient Greeks seem to have celebrated their power over the environment and to have seen man as "the orderer of nature."[1] The poem's distinctive idiom of supererogation, of gargantuan scale and limitlessness, was thus much admired by ancient literary critics. It was regarded by "Longinus" as lending it true sublimity or elevation. Homer wins the *Contest of Homer and Hesiod* because his evocation of improbably enormous quantities and distances enhances his poetry's glamorous impact.

The Homeric *Iliad*, an epic poem now twenty-eight centuries old, has recently reached wider audiences and penetrated to deeper levels of popular culture than ever before, through cinema and TV adaptations, feminist reappraisals, fiction, poetry, and computer games. In the twenty-first century, at the same time as the idea of the "Anthropocene" has made a mark on the public imagination, and some geologists are dating its inception to the Bronze Age, creative artists have just begun to respond to the sense of excessive consumption and ensuing apocalypse that underlies the epic poem. But scholarship has lagged behind.

Over the past few decades, new initiatives in ecological thinking about human literary culture have emerged, initiatives that are beginning to be acknowledged and implemented by classicists as well. These approaches have been fused in this book with both a more old-fashioned aesthetic appreciation and an interest in the relations between the humans performing and profiting from the labor required of them. This reveals the *Iliad*'s absolute erasure of mining and transporting ore to smithies, and meager emphasis on hard domestic or agricultural drudgery. Men die by the hundreds in the poem, but for the real-world counterparts of those who fell in battle, it was relentless labor that rendered their lives, in Thomas Hobbes's words, "poor, nasty, brutish, and short."[2] The bifurcated approach, which blends ecological thinking with Marxist interest in human modes of production and the labor they entailed, has suggested some of the questions asked in this book about the *Iliad*'s "political ecology" and "environmental unconscious," as well as the methods used to try to answer them.

Planet Earth has existed for 4.5 billion years and life for about 3.5 billion. Modern hominids appeared only a half million years ago, but people as we know them emerged only between about 100,000 and 70,000 years before our era. The *Iliad* shows that the seeds of environmental catastrophe were already sown by warfare at the dawn of human civilizations less than ten thousand years ago. Reading the *Iliad* ecologically gives us a new interpretation to add to the previous shifts in understanding and repurposing it across time. But my objective is also to encourage the epic's readers to take action in our current battle for the survival of our planet's ecosystems. We need Achilles to go green and make do with just one tripod hereafter.

Marlene Sokolon has argued that we can repurpose the *Iliad* for our turbulent times by reinterpreting it as an early example of protest poetry. We can leverage Achilles's "challenge to authority, anger at injustice, and confrontation with the fragility of the human condition." The *Iliad* "provides insights into why human beings protest, connects political poetry to philosophic questions, and highlights the human being as a perennial protester who must face the inevitable choice of safety or perilous political action."[3] Emily Katz Anhalt, similarly, believes that retelling the story of Achilles's rage can help us to "see the costs of rage and violent revenge and to cultivate more constructive ways of interacting."[4] Sokolon and Katz Anhalt are not thinking specifically about protesting the failure of our rulers and industrialists to address continuing human depredation on the environment, but their points are well taken.

The war in Ukraine, which began in 2022, and the Israel-Hamas conflict that exploded in October 2023, have retarded international cooperative initiatives aimed at reversing climate change. They have also already caused many thousands of deaths. Journalists and poets have inevitably begun to draw parallels between both wars and the tragic conflict portrayed in the greatest classical war poem, the Homeric *Iliad*.[5] As we watch the bombardment of Ukrainian and Middle Eastern cities, and the terror of women and children in both arenas of combat, the tears of the widowed Andromache in the *Iliad* feel to many of us more agonizingly relevant than at any time since World War II.

The poem not only foresees innumerable fatalities and the annihilation of Trojan civilization. It also predicts the total depopulation of what is now northwestern Turkey, and describes the aggressive and irreversible Bronze Age deforestation undertaken to provide the vast amounts of timber and metal needed to support both naval and land militias and to clear land for arable farming and livestock grazing, cooking, and sacrifice. The scale of the tree-felling in the *Iliad* must prompt us to ask, with the environmentalist and philosopher Christopher Stone, whether trees need to have legal protection and standing.[6] There has always been cavalier abuse of natural resources by humans in conflict.[7] Within a few weeks, the Russian invasion of 2022 had destroyed substantial parts of the Ukrainian infrastructure and poisoned its rivers and forests. In the arid Middle East, war is exacerbating already urgent challenges presented by rising sea levels, desertification, and declining precipitation.

A Trojan fir on
Mount Ida, now
the Kazdağı
National Park.
(Photograph taken
in 2005 by Volker
Höhfeld. https://
creativecommons
.org/licenses/by
-sa/4.0/deed.en)

The terrible immediacy of these wars has, disastrously, pushed the
environmental crisis much further down humanity's list of priorities. But
these conflicts are also directly and immeasurably exacerbating pollution.
The *Washington Post* has reported on the "untold volumes of toxins and pol-
lutants" recently released into the atmosphere in Ukraine.[8] Cluster bombs
and thermobaric rockets have ignited massive clouds of poisonous aero-
sols. Increased wildfire risk and biodiversity loss will blight Ukraine and
the Middle East for generations. The pollution of water sources in and
around the Gaza Strip will take years, if not decades, to reverse.[9]

To end, however, on a cautiously optimistic note: The trees of Mount
Ida have recently been protected by protestors—at least temporarily—from

devastation by the contemporary mining industry. Part of the area was declared to be a national park in 1993, but the Turkish state subsequently sold land and mining rights for an enormous sum to the Canadian mining company Alamos Gold Inc. The proposed mining project is just twenty kilometers from Troy. In 2017 the Turkish project partner, Doğu Biga, began felling thousands of trees and removing all of the soil, down to bare rock. Around 200,000 trees were cut down. Cyanides began to be used for gold extraction, putting drinking water supplies at risk.[10] But the extent of the tree clearance and destruction of natural environments was detected by satellites and drones. Images were collected by a Turkish environmental organization, and a large protest camp was set up in 2019. Operations were successfully stalled, and the Turkish government removed Alamos's mining licenses. The company has responded by registering a claim against the Republic of Turkey with the International Centre for Settlement of Investment Disputes.[11]

People's action can work. It has, for the time being, saved some of the last remaining forests of Mount Ida, where the wood, including the Trojan fir or *Abies nordmanniana* ssp. *equi-trojani*, never has been infinite—regardless of what Homer's heroes say. By accessing the *Iliad*'s ecological unconscious, now more than three millennia old, we can, as humans, enrich our struggle to ensure a better future. The *Iliad* is not only the poem *of* the Anthropocene; it has the potential truly to become the epic of the Earth—the poem *for* the Anthropocene.[12]

Acknowledgments

My interest in trees was first stimulated by my paternal aunt, Jean Hall, an expert botanist who specialized in plants of the eastern Mediterranean and paid for me to adopt my own broad-leaved tree in the protected Eoves Wood in Hampshire. For this book, I received timely and abundant advice on the whole project from Matt Shipton, Jason König, Christopher Schliephake, and Brooke Holmes. David Braund talked to me about trees in the ancient world on research trips to the Black Sea and across Bulgaria; Boris Rankov patiently responded to questions about timber in the construction of ships; in conversations with Emily Wilson, she encouragingly shared her appreciation of the importance of timber similes in the *Iliad;* Patrice Rankine and Sara Monoson made enlightening suggestions about relevant reading; supervisions with my Durham PhD student Emma Bentley have also been enlightening.

Becky Brewis is responsible for the gorgeous line drawings; Lucia Kustra, a George Washington University Archaeology student, took the photograph of the Homeric litigation scene on the doors of the Supreme Court, generously facilitated by Eric Cline. Nat Haynes, Jo Balmer, Marina Carr, and Alicia Stallings allowed me to reproduce passages of or about their stunning creative responses to the *Iliad*. Heartfelt thanks go to Katrina Kelly for suggesting the resonant title. Various versions of some of the chapters have been delivered at several venues where I was warmly welcomed and received invaluable feedback: I am particularly grateful to Vinnie Nørskov at Aarhus University, Denmark; the classics postgraduates at the University of Washington, Seattle; the Liverpool branch of the Classical Association, UK; Maria Fernanda Brasete at Aveiro University, Portugal; the Classics Department at Vilnius University, Lithuania; Bristol Royal Grammar School; the British Academy; Justin Vyvyan-Jones at the Oxford University Classics Society; the Saint Andrews Centre for Environmental Studies; staff and students at Brandeis, Wesleyan, and Brown Universities; and Christine Walde at the Johannes Gutenberg University in Mainz, Germany.

George Gazis gave me crucial feedback at a Durham University seminar. At a conference at the Academy of Athens, Tim Whitmarsh suggested gently that I might need to think more about the gods; the comments of both Simon Goldhill and Richard Janko were also invaluable. Paul Cartledge gave freely of his time, reading a draft of the entire book and providing generous and indispensable comments. Nosheena Jabeen expertly solved a bibliography crisis at the last minute, with good cheer and swiftly. At Yale University Press, Heather Gold has consistently supported the project, Julie Carlson's copy-editing has been kind and conscientious, Jeffrey Schier steered the book ably and amiably through production, Meridith Murray produced a meticulous index, and the anonymous reviewer's comments were exceptionally helpful. But, as always, my greatest debt is to my husband, Richard Poynder, who loves trees as much as I do.

Notes

All translations in the book are mine unless otherwise indicated.

Prologue

1. Hegel, *Vorlesungen*, 5.29.
2. See Marrou, *History of Education*, 162, and Pack, *Greek and Latin Literary Texts*, nos. 2707, 1208.
3. Hall, "Classics Invented," 36.
4. Ruskin, "Mystery of Life," 109.
5. Graziosi and Greenwood, *Homer in the Twentieth Century*; Hall, *Return of Ulysses*; McConnell, *Black Odysseys*.
6. Rudd, *Greenery*, 35.
7. Ghosh, *Great Derangement*, esp. 3–84; see also König, *Folds of Olympus*, xxiv.
8. *Oxford English Dictionary* s.v. "apocalypse," draft addition b. https://www.oed.com/dictionary/apocalypse_n#131062527.
9. On apocalyptic and postapocalyptic fiction, see the excellent studies by Tate, *Apocalyptic Fiction*, and Payne, *Flowers of Time*, respectively.
10. E.g., Wallace-Wells, "Uninhabitable Earth."
11. Colebrook, *Death of the Posthuman*.
12. Abram, *Evergreen Ash*, 17.
13. According to Lewis and Maslin, "Defining the Anthropocene"; see also Parham and Westling, *Global History*, 6.
14. Crutzen and Stoermer, "Anthropocene."
15. Schliephake, *Environmental Humanities*, 2–3.
16. Subramanian, "Humans versus Earth."
17. Smith and Zeder, "Onset of the Anthropocene."
18. Parham and Westling, "Introduction," 5.
19. Haber, "Energy," and "Anthropozän."
20. Morris, *Foragers*, 65.

21. Morris, *Foragers*, 65.
22. Chew, *World Ecological Degradation*, 171–73; Hern, "Is Human Culture Carcinogenic?"
23. Habermas, *Legitimation Crisis*, 15.
24. Lewis and Maslin, "Defining the Anthropocene," 171; Parham and Westling, "Introduction," 8.
25. Szerszynski, "End of the End," 171.
26. Thirgood, *Man*.
27. Headrick, *Humans*, 93.
28. Perlin, *Forest Journey*, 60.
29. On questioning deforestation see, notably, Rackham, "Ecology." For more on skepticism related to humans' negative effect on the environment, see Middleton, "Nothing Lasts," esp. 31–33, and McAnany and Yofee, *Questioning Collapse*.
30. Williams, *Deforesting the Earth*, 95.
31. Liritzis, Westra, and Miao, "Disaster Geoarchaeology." Terence Hawkins adapts the *Iliad* to a modern context in his *Rage of Achilles*, a racily written novel informed by Julian Jaynes's theory of the bicameral mind.
32. Parham and Westling, "Introduction."
33. Parham and Westling, "Introduction," 9.
34. Parham and Westling, "Introduction," 10.
35. Walcott, *Omeros*, 1–6, 8, 31, 47, 59. I owe this reference to Patrice Rankine.
36. Williams, *Country and the City*.
37. Holmes, "Foreword," xi.

Chapter One. Changing Interpretations of the *Iliad*

Epigraph: Homer, *Iliad*, 1.1–5.
1. Scholiast D on *Iliad* 1.5. I have slightly adapted the version of West, *Greek Epic Fragments*, 80–83.
2. Hesiod, *Theogony. Works and Days. Testimonia*, 21.
3. *Cypria* fragment 1, which I have slightly adapted from the translation of West, *Greek Epic Fragments*, 82–83. For the original *Cypria* fragment, see https://www.theoi.com/Text/EpicCycle.html.
4. Slightly adapted translation by Most, *Hesiod* (his fragment no. 155), 256–59 of supplemented papyrus P. Berol. 10560 95–16. For more on this papyrus, see https://berlpap.smb.museum/record/?result=1&Alle=10560&lang=en.
5. Perlin, *Forest Journey*, 8.
6. Rappenglück and Rappenglück, "Does the Myth."
7. I assume, however, that multiple authors, over many years, decades, and even centuries, contributed to the development of the poems that have come down to us under the names of both "Homer" and "Hesiod." See Kidd, "Inventing the 'Pygmy,'" and Dan, "Mythic Geography."

8. For the Telchines, see Strabo, *Geography*, 14.2.7. This is viewable in translation along with other sources on these mythical craftsmen at https://www.theoi.com/Pontios/Telkhines.html.

9. Janko, "Summary," 588.

10. Ovid, *Metamorphoses*, 7.365–68; Pindar, *Paean* 5; Nonnus, *Dionysiaca*, 18.35; Servius, *Servii grammatici qui feruntur in Vergilii carmina commentarii*, 4.377.

11. Liritzis, Westra, and Miao, "Disaster Geoarchaeology," 1317–19. For discussions of the use of antiquity in defining the concept of "collapse," see Middleton, "Nothing Lasts" and *Understanding Collapse*.

12. Ovid, *Metamorphoses*, 1.313–47; see also Griffin, "Ovid's Universal Flood."

13. See, e.g., Frost, "*Critias.*"

14. For more on Plato's *Atlantis*, listen to commentary by Christopher Hill on the BBC series *In Our Time*, https://www.bbc.co.uk/programmes/m001c6t3; see also Vidal-Naquet, *Atlantis Story*; and Hall, "Nine Thousand Years Ago."

15. Thucydides, *History of the Peloponnesian War*, 2.49–51; quotation at 2.48.

16. For the sources see Daugherty, "Cohortes Vigilium," and, on the consequences, Newbold, "Social and Economic Consequences."

17. Hall and Stead, *People's History*, 123–25; Simmons, "Bulwer"; Moormann, *Pompeii's Ashes*.

18. *Bibliotheca Historica*, 1.47.

19. Shelley, *Complete Poetical Works*, 605.

20. Hall, "Ozymandias."

21. The episode is the fourteenth of the fifth season. It was written by Moira Walley-Beckett, who had long wanted to use the poem, and directed by Rian Johnson. It first aired on AMC in the United States and Canada on September 15, 2013.

22. Oppenheimer was filmed on July 16, 1945, quoting Krishna's statement in the *Bhagavad Gita* 11.32. https://www.youtube.com/watch?v=lb13ynu3Iac.

23. Lisboa, *End of the World*, 105.

24. Sedia, "Shelley's 'Ozymandias.'"

25. Shelley, *Complete Poetical Works*, 320.

26. See Walter, "Poseidon's Wrath"; Borsch and Carrara, "Zwischen Natur."

27. γαῖαν ἀπειρεσίην (20.58).

28. Osborn, *Limits*, 17.

29. ἀπὸ χαλκοῦ θεσπεσίοιο.

30. πλείστους Φρύγας (3.185); İhlas News Agency, "Pollution."

31. Kirk, *Iliad*, vol. 2, 216, on lines 6.425–28.

32. ὅσα ψάμαθός τε κόνις τε (9.381–85); see Kirk, *Iliad*, vol. 1, 245, on line 2.800. Leaves are usually symbols of renewability rather than multitude; for sand in this latter sense, see 9.385.

33. μυρία φῦλα περικτιόνων ἐπικούρων (17.220).

34. ὀκτάκνημα (5.723). See Kirk, *Iliad*, vol. 2, 133, on lines 5.722–23; Lorimer, *Homer*, 319.

35. χρυσείην, ἑκατὸν πολίων πρυλέεσσ᾽ ἀραρυῖαν (5.744). See Kirk, *Iliad*, vol. 2, 135, on lines 5.743–44.
36. Kirk, *Iliad*, vol. 2, 53, on line 5.4.
37. οὐδ᾽ εἴ κε τὰ νείατα πείραθ᾽ ἵκηαι / γαίης καὶ πόντοιο (8.478–779); πολυφόρβου πείρατα γαίης (14.200). See Bergren, *Etymology*, 21, 106–7 and 111 on *peirar* referring to the ends of the Earth, the limit of the human world, in early Greek poetry, including *Iliad* 14.200–201, 14.301–2. See also the Homeric *Hymn to Aphrodite* 226–27; Hesiod, *Works and Days*, 168, and *Theogony*, 333–35, 622, 738, 809. On the formidable powers with which Ocean is invested in the *Iliad*, albeit "with the lightest of touches," see Ali, "Oceanos," 241–42.
38. Heiden, *Homer's Cosmic Fabrication*, 187; Hall, *Theatrical Cast*, 344–49; Hall, "One Precise Day."
39. Longinus, *On the Sublime*, 16, 9; this is a conflation of two Iliadic passages: 21.388 (confused in the quotation with 5.750), and 20.61–65.
40. Longinus, *On the Sublime*, 9.
41. Longinus, *On the Sublime* 15; *Iliad*, 20.170; Lonsdale, *Creatures of Speech*, 1.
42. Dineen, "Catastrophic Libyan Dam."
43. See, e.g., Floyd, "*Kleos Aphthiton*"; West, *East Face, Indo-European Poetry*, and "Homeric Question"; Haubold, *Greece and Mesopotamia*.
44. See further Wilson, "Gilgamesh Epic"; West, *East Face*, 344–46.
45. West, *East Face*, 336–38.
46. Haubold, *Greece and Mesopotamia*, esp. 36–51, 64–70.
47. This exciting story is well told in Damrosch, *Buried Book*; see also Hall, "George Smith," on the working-class autodidact and intellectual titan, whose translations in the 1870s brought the world the Babylonian creation poem *Enuma Elish* and the *Epic of Gilgamesh*, which Smith miraculously deciphered from clay tablets discovered by the Assyrian Hormuzd Rassam in 1853.
48. Hall, *Return of Ulysses*.
49. On the exceptions, see Hall, "Classical Epic"; Hall and Stead, *People's History*, 3–6.
50. Foerster, "Homer, Milton," 76–77.
51. Drummond, "Battle of Trafalgar," 31.
52. Gladstone, *Studies*, vol. 1, 24, 87, 162, 174, 182, 317 n.15, 533.
53. See Rossetti, *Notes*, 27; Stephens, "Fine Arts."
54. Dunn, "Common Man's 'Iliad.'"
55. Pease, "Fear, Rage," explores the significance of anger in the novel.
56. Dunn, "Common Man's 'Iliad,'" 277.
57. See Hall and Macintosh, *Greek Tragedy*, 509–10.
58. This prescient tragedy was recently revived, in a production of Nina Murray's English translation, *Cassandra*, directed by Helen Eastman and sponsored by the Ukrainian Institute, London. It was aimed at raising awareness of the suffering endured by Ukrainians since the Russian invasion of 2022.

59. Vandiver, *Stand in the Trench*. See also Dué, "Learning Lessons."

60. On Shaw-Stewart and this extraordinary poem, see the website maintained by Balliol College, Oxford: https://archives.balliol.ox.ac.uk/Past%20members/PHStewart.asp#gsc.tab=0.

61. Jones, *In Parenthesis*, 160.

62. Hall and Stead, *People's History*, 497–511.

63. Weil, *Simone Weil's "The Iliad*," 3. Weil's essay was first published in 1940 in *Cahiers du Sud* as "*L'Iliade*, ou le poème de la force." The English translation by Mary McCarthy was first published in the November 1945 issue of *Politics*. See further Gold, "Simone Weil."

64. Benfey, "Introduction," vii–xxiii.

65. Schadewaldt, *Iliasstudien*, 108.

66. Wenders, *Der Himmel über Berlin*, at 37:11 to 38:37 minutes.

67. Wenders, *Der Himmel über Berlin*, at 38:26 minutes.

68. Harrison, *Collected Poems*, 317.

69. Harrison, *Gaze of the Gorgon* at 00:39 minutes; the film can be viewed by appointment at the Archive of Performances of Greek & Roman Drama, of which I was a co-founder, at Oxford University. Weil, *Simone Weil's "The Iliad*," 26.

70. Harrison, *Collected Film Poetry*, 166. See also the discussion in Stead, "Only Tone for Terror."

71. Harrison, *Collected Film Poetry*, 168. The film and its relationship with the *Iliad* are discussed in detail in Hall, *Tony Harrison*, 125–29.

72. Harrison, *Gaze of the Gorgon*, at 45:44 minutes.

73. Harrison, *Collected Film Poetry*, 170–71.

74. Longley, *Ghost Orchid*, 39.

75. Shay, *Achilles in Vietnam*.

76. Wahl, "Préface," 7, translated in Schein, "Reading Homer," 19.

77. Hall, *Return of Ulysses*.

78. Hall and Stead, *People's History*, 4–5, 48–50. On the history of the reception of the *Iliad*, see also Martin, "Introduction," 53–64.

79. Cole, "Think Again," 30.

80. Cole, "Think Again," 32.

81. E.g., Tobin, *Siege*; Manfredi, *Talisman*; Clarke, *War at Troy*; Pavlou, *Gene*; Elyot, *Memoirs*; Gemmell, *Troy: Lord of the Silver Bow*, *Troy: Shield of Thunder*, and *Troy: Fall of Kings*; George, *Helen of Troy*.

82. On the interest in Euripides's work after 9/11, see Hall, "Iphigenia and Her Mother."

83. Hall, "Greek Tragedy."

84. Personal correspondence with Unsworth in 2003.

85. It is a shame that the contemporary political resonances of *Troy* receive such sparse attention in Winkler, *Troy*, an otherwise excellent collection of essays.

86. *PR Newswire*, "White House."

87. For Miller's bellicose politics, hear his comments during an edition of National Public Radio's "Talk of the Nation" on January 24, 2007, at https://www.npr.org/2007/01/24/7002481/writers-artists-describe-state-of-the-union. For more on the film, see Kovacs, "Truth," 391–92.

88. Stax, "Stax Report."

89. Walsh, "Warrior." See some further statements along the same lines by Petersen, quoted in Winkler, *Troy*, 7–8.

90. Viewable online at https://www.youtube.com/watch?v=-dQSfAoEWxE&feature=youtu.be&fbclid=IwARoOK2zRm1T-au3a5mL349yERehkuYswOEdP-OCFfA1peSuvlDxGLrfhz.

91. Thomas, Rivera, and Sepulveda, *Marvel Illustrated*.

92. Produced by the Italian record label Underground Symphony.

93. Published by Sega as the second installment in the *Total War Saga* subseries.

94. *Ransom* was shortlisted for the 2011 International IMPAC Dublin Literary Award and received the 2009 John D. Criticos Prize.

95. Malouf, *Neighbours*, 13, and McNeill, "Classics Today"; see also Reynolds, "Poem of Force."

96. Weil, *Simone Weil's "The Iliad,"* 31.

97. Malouf, *Ransom*, 216.

98. In Balmer, *Chasing Catullus*, 41–42.

99. See Iraq Body Count, "Iraqi Deaths."

100. Hahnemann, "Book of Paper."

101. Oswald, *Memorial*, 66–67; see also *Iliad*, 21.17–26.

102. Oswald, *Memorial*, 1. See, e.g., Demetrius, *On Style*, 320.

103. Oswald, *Memorial*, 2.

104. Oswald, *Memorial*, 19.

105. Oswald, *Memorial*, 13.

106. Oswald, *Memorial*, 69.

107. Oswald, *Memorial*, 37.

108. Oswald, *Memorial*, 33.

109. Oswald, *Memorial*, 26.

110. Warner in Warner et al., "Symposium on Crying," 18.

111. See Wyles, "Ménage's Learned Ladies." Wilson's translation, like her brilliant *Odyssey*, is a landmark in Homer studies.

112. Hall, *Return of Ulysses*, 116–18; Bentley, quoted in Grote, *History of Greece*, vol. 1, 151n.

113. William Golding, in conversation, quoted in Boitani, *Shadow*, viii.

114. Hughes's book won the London Hellenic Prize in 2018.

115. Keating, "Marina Carr."

116. Hall, "Why Is Penelope."

117. The literary prizes were the Women's Prize for Fiction 2019 and the Costa Novel Award.

118. As explored in Timberlake Wertenbaker's version *Our Ajax*, which premiered at the Southwark Playhouse in London in November 2014.

119. Barker, *Silence*, 39; see also 212–13.

120. Barker, *Silence*, 19.

121. See Lanone, "L'état d'exception."

122. Longinus, *On the Sublime*, 9.7.

123. Haynes, *Thousand Ships*, 339.

124. Haynes, *Thousand Ships*, 55.

125. Haynes, *Thousand Ships*, 276–77.

126. Stallings, "Crown Shyness."

127. Hall, *Aeschylus' Agamemnon*, 677–78.

128. Euripides, *Trojan Women*; for the closing dirge, see 1297–99; for "come to nothing," see 1240–42. For more on this topic, see Hall, "Trojan Suffering."

129. West, *East Face*, 378.

130. Hall, "Can the *Odyssey*."

131. Plato, *Ion*, 535b–c.

132. Oswald, *Memorial*, 72.

133. Oswald, *Memorial*, 84.

134. Barker, *Silence*, 322–23.

Chapter Two. The *Iliad* in Its Historical Contexts

1. Thucydides, *History*, 3.98, 4.29–30, 34; Thommen, *Environmental History*, 40.

2. König, *Folds of Olympus*, 39.

3. Tomaselli, "Degradation"; Williams, *Deforesting*, 77.

4. Oppenheimer and Pyle, "Volcanoes," 450.

5. Oppenheimer and Pyle, "Volcanoes," 452.

6. Papadopoulos, "Tsunamis," 499; Heiken and McCoy, "Caldera Development."

7. Oppenheimer and Pyle, "Volcanoes," 454 with further bibliography.

8. Papadopoulos, "Tsunamis," 499–500, 510; S. Solovieva, O. Solovieva, Go, Kim, and Shchetnikov, *Tsunamis*.

9. Stiros, "Earthquakes," 476–77.

10. Stiros, "Social and Historical Impacts."

11. Lloret, Pinol, and Castellnou, "Wildfires," 541; Roberts and Reed, "Lakes."

12. Thornes, "Land Degradation," 563.

13. See Naveh and Dan, "Human Degradation."

14. Thornes, "Land Degradation," 563.

15. del Carmen Llasat, "Storms," 514–15.

16. Maas and Macklin, "Impact"; Thorndycraft, Benito, Barriendos, and Llasat, *Palaeofloods*; del Carmen Llasat, "Storms," 529.

17. Pindar, *Paean*, 4.35–54.

18. Janko, "Summary," 587–88.

19. See Clay, *Hesiod's Cosmos*, 91 and n.11.

20. See Mayor, *Fossil Hunters.*
21. Fletchman Smith, *Mental Slavery*, 8–9; Hall, "Playing Ball," 204–5.
22. Schaeffer, *Stratigraphie.*
23. E.g., Walløe, "Disruption."
24. Carpenter, "Discontinuity."
25. Drake, "Influence."
26. Kaniewski, Paulissen, Van Campo, Weiss, Otto, Bretschneider, and Van Lerberghe, "Late Second–Early First Millennium."
27. Perlin, *Forest Journey*, 49–50.
28. Dimopoulou, "Metallurgy," 135–41.
29. Perlin, *Forest Journey*, 51.
30. Perlin, *Forest Journey*, 52–54.
31. Knapp and Manning, "Crisis," 126.
32. Dickinson, "Parallels," 136–37.
33. Levi, *Atlas*, 44–49.
34. *Iliad* 3.271, 19.252; see also Lang, "Bronze and Iron," 289; Tsountas and Manatt, *Mycenaean Age*, 204.
35. Janko, "Helen of Troy," 121.
36. σήματα λυγρά,/ γράψας ἐν πίνακι πτυκτῷ θυμοφθόρα πολλά.
37. Jahns, "Untersuchungen." See also Thommen, *Environmental History*, 37.
38. Knapp and Manning, "Crisis," 193.
39. Cline, *1177 B.C.*, 37; Mylonas, *Mycenae*, 213–33; Sherratt, "Reading the Text."
40. West, "Rise of the Greek Epic," 158.
41. Lonsdale, *Creatures of Speech*, 3.
42. Ventris and Chadwick, *Documents*, 94–105.
43. Ventris and Chadwick, *Documents*, 183–84.
44. Ventris and Chadwick, *Documents*, 145, 159.
45. Palaima, "Metaphysical Mind," 483–84.
46. Ventris and Chadwick, *Documents*, 125–29.
47. Ventris and Chadwick, *Documents*, 128; see also Palaima, "Metaphysical Mind," 480.
48. Ventris and Chadwick, *Documents*, 158–59.
49. Ventris and Chadwick, *Documents*, 123.
50. Ventris and Chadwick, *Documents*, 135.
51. Ventris and Chadwick, *Documents*, 132. On "real-world" human-animal relationships in the Aegean middle and late Bronze Age, see Laffineur and Palaima, *Zoia.*
52. Cline, *1177 B.C.*, 37.
53. Sackett and Popham, "Lefkandi."
54. Krapf, "Symbolic Value," 531–32; on similar heirlooms in Laconia, see Janko, "Summary," 592.
55. Desborough, Nicholls, and Popham, "Euboean Centaur."
56. West, "Rise of the Greek Epic," 159–65.

57. Diodorus Siculus, *Diodorus of Sicily*, 1.8; Bosak-Schroeder, *Other Natures*, 23.

58. Aristotle, *Politics*, 1.1256a–b; for Dicaearchus of Messana, see Bosak-Schroeder, *Other Natures*, 24.

59. Payne, *Flowers of Time*; Möller, "Experience"; on the myth of the successive ages or races of man, see also Baldry, "Hesiod's Five Ages."

60. Hesiod, *Works and Days*, 145–46; οἷσιν Ἄρηος/ἔργ᾽ ἔμελεν στονόεντα καὶ ὕβριες.

61. Hesiod, *Works and Days*, 150, 159–60; ἀνδρῶν ἡρώων θεῖον γένος, οἳ καλέονται/ἡμίθεοι, προτέρη γενεὴ κατ᾽ ἀπείρονα γαῖαν.

62. Hesiod, *Works and Days*, 156–73.

63. Hesiod, *Works and Days*, 174–99.

64. On the exceptions, Pandarus's arrowhead and Areithoüs's mace (4.123, 7.141), see Lang, "Bronze and Iron," 282–83.

65. Hesiod, *Works and Days*, 646–62.

66. Herodotus, *Histories*, 1.142; Hall, *Aeschylus' Agamemnon*, 1257; Aeschylus, *Suppliant Women*, in Aeschylus, *Persians. Seven against Thebes. Suppliants. Prometheus Bound*, 686.

67. When it comes to place, the Troad of the *Iliad*, and the topographically identifiable, Ithacan and Peloponnesian parts of the *Odyssey*, are, according to the World Wildlife Fund classification system, at least all set in a single ecoregion. What classicists know as the world of Homer is, botanically, Ecoregion #PA1201, the "Aegean and Western Turkey Sclerophyllous and Mixed Forests Ecoregion." This in turn is one constituent of Biome #12, "Mediterranean Forests, Woodlands, and Scrub." Head, *Deep Agroecology*, 304, 306.

Chapter Three. Defining an Ecocritical Approach

1. König, *Folds of Olympus*, 36.

2. König, *Folds of Olympus*, 46.

3. Schama, *Landscape*, 95; see also 140–58 and Nardizzi, *Wooden Os*.

4. König, *Folds of Olympus*, xxiv and 36; Heise, *Sense of Place*, 76–77 on such phenomena in modernist literature; Kerridge, "Ecocritical Approaches," 369.

5. Head, *Deep Agroecology*.

6. Head, *Deep Agroecology*, "Foreword."

7. Head, *Deep Agroecology*, 72.

8. Head, *Deep Agroecology*, 160. Head is more interested in the *Iliad* than the *Odyssey*, although he claims that Penelope "might usefully be regarded metaphorically as "the natural world," or more specifically as "the force of vibrant equilibrium in that natural world, the special chemistry of interdependence that keeps the systems operating" (348).

9. Head, *Deep Agroecology*, 447.

10. Head, *Deep Agroecology*, 468.

11. Gee, *Short History*, esp. 212, 228.

12. Rudd, *Greenery*, 33.

13. Martin, *Language of Heroes.*

14. See Heilbrun, "What Was Penelope"; Hall, *Return of Ulysses*, 118–29.

15. See, e.g., Greenwood, *Afro-Greeks.*

16. See, e.g., Clarke, "Queer Classics"; Olsen and Telò, *Queer Euripides.*

17. Shipton, *Politics of Youth.*

18. Hall and Stead, *People's History;* Hall, *Tony Harrison.*

19. Hall, "Immortal Forgotten Other Gang."

20. Quoted in Lahr, "Inventing the Enemy."

21. Taplin, *Greek Fire.*

22. Walcott, *Epitaph*, 15; see also Walcott, *Omeros*, 271.

23. Hughes, "Artemis," 193.

24. Usher, *Plato's Pigs*, xi.

25. Siewers, "Ecopoetics."

26. Vidal-Naquet, "Oedipus."

27. Williams, *Country and the City*, 120–27.

28. Bakhtin, "Response."

29. See his essay "Figura" (1944) as translated by Ralph Manheim in Auerbach, *Scenes*, 11–76.

30. Cohen and Duckert, *Elemental Ecocriticism*, 5.

31. Ghosh, *Great Derangement.*

32. Cohen and Duckert, *Elemental Ecocriticism*, 6.

33. O'Connor, *Natural Causes*, 54.

34. See Hall, "Karl Marx."

35. The term "dialectical materialism" was coined in 1887 by Joseph Dietzgen, a correspondent of Karl Marx, although Engels had already used the formulation "materialist dialectic" in 1883. Marx himself had used the phrase "materialist conception of history," which was later expressed as "historical materialism" by Engels.

36. McInnes, *Western Marxists*, 43, and Illingworth, *Divine Immanence*, 137.

37. Boscagli, *Stuff Theory*, 17–18, and the anti-Marxist new materialists critiqued by Wilkie, "Introduction."

38. Quoted in Khan, "Vital Materiality," 46.

39. Khan, "Vital Materiality," 46.

40. See Hall, *Aristotle's Way*, ch. 8.

41. Hall, "Materialisms."

42. Data extracted from https://worldpopulationreview.com/country-rankings /employment-in-agriculture-by-country.

43. Callinicos, "New Middle Class."

44. Wright, *Life of Joseph Wright*, 1:131, 189; Hall, "Classics in Our Ancestors' Communities"; Hall and Stead, *People's History*, 303.

45. Hunter, "Gothic Psychology"; see further Devall and Sessions, "Deep Ecology."

46. Hunter, "Gothic Psychology"; Mathews, *Ecological Self*, 3.

47. Naess, "Self-Realization," 227.

48. Rowland, *Ecocritical Psyche*, viii.

49. Roszak, *Voice of the Earth*, 107–8; Roszak, "Awakening."

50. The first version was published in German in 1935, and is reproduced in English translation, edited by Hannah Arendt, in Benjamin, *Illuminations*.

51. Gates, *Figures in Black*, 50.

52. Guzowska, Becks, and Andersson-Strand, "She Was Weaving."

53. On Mycenaean ivory production, see Luján and Bernabé, "Ivory and Horn Production."

54. Karavites, *Promise-Giving*, 3.

55. Rose, *Sons of the Gods*, 43–91.

56. Gottschall, "Homer's Human Animal," 284.

57. Rose, "Homer's *Iliad*."

58. Brockliss, *Homeric Imagery*.

59. Heise, "Hitchhiker's Guide," 503.

60. Chakrabarty, "Climate."

61. Haraway, *When Species Meet*, 15–16, 25, 32, 62, 98, 113 and 138; Holmes, "Situating Scamander."

62. See Heise, "Hitchhiker's Guide," and, on the institutional formation of ecocriticism, Garrard, "Introduction."

63. Foucault, *Birth of Biopolitics*.

64. Medovoi, "Biopolitical Unconscious," 125–26.

65. Clark, *Ecocriticism*, esp. 9–24.

66. Farina, *Ecosemiotic Landscapes*.

67. Siewers, *Re-Imagining*, 27.

68. Schliephake, *Environmental Humanities*, 7. The first issue of the *Journal of Environmental Humanities* came out in 2012.

69. Schliephake, *Environmental Humanities*, 5; Homer, *Odyssey*, 8.109–18.

70. Garrard, "Introduction," 3, 5.

71. E.g., Bergthaller and Mortensen, *Framing*; Siewers, *Re-Imagining*.

72. Schliephake, *Environmental Humanities*, 57. The current book is in part a response to this plea.

73. Schliephake, *Ecocriticism*, 10; Hutchins, "Interspecies Ethics"; Chinn, "Ecological Highway"; Gifford, "Environmental Humanities."

74. Schliephake and Zemanek, *Anticipatory Environmental (Hi)Stories*, 3. See also Backman and Cimino, *Biopolitics and Ancient Thought*.

75. Perlin, *Forest Journey*, 44–45.

76. Quoted in Meiggs, *Trees and Timber*, 378.

77. Dalley, "Natural World," 22.

78. Dalley, "Natural World," 24. It was the king's prerogative in reality to kill lions.

79. Dalley, "Natural World," 21, 23; Pausanias, *Guide to Greece*, 8.29.3–4.

80. Dalley, "Natural World," 33.
81. Dalley, "Natural World," 22–23.
82. Dalley, "Natural World," 30.
83. Dalley, "Natural World," 28–29.
84. Siewers, "Ecopoetics"; White, "Historical Roots."
85. Green, "Garden of Eden," 53. See also 61 for her remarks on the cedars of Lebanon, which once were said to have been more beautiful even than the trees of Eden (Ezekiel 31:6–9). The biblical reference is to Genesis 1.28 (KJV).
86. Thornber, "Environments," 39.
87. Thornber, "Environments," 39–40.
88. Sivaramakrishnan, "Ecopoetics," 75–76.
89. Christenson, "Who Shall Be."
90. Tlili, "I Invoke God."
91. Abram, *Evergreen Ash.*
92. Rudd, *Greenery*, 81–83.
93. Holmes, "Situating Scamander."
94. Eckerman, "Ancient Greek Literature."
95. Cook, "Remaking Eighteenth-Century Ecologies."
96. Palaima, "Metaphysical Mind," 482–83. The scepter is described in language with outstandingly clear parallels in Linear B.
97. Nagy, *Best of the Achaeans*, 183–84.
98. Gee, *Short History*, 228.

Chapter Four. Nature and the Divine

1. εὐόμενοι τανύοντο διὰ φλογὸς Ἡφαίστοιο (23.33); ὑπείρεχον Ἡφαίστοιο (2.426).
2. Veldsman, "Place of Metaphysics," 7.
3. Trépanier, "Early Greek Theology," 275.
4. Karavites, *Promise-Giving*, 4.
5. πορφυρέην ἶριν (17.548–52).
6. Marinatos, "Myth," 3–10.
7. Vetters, "All the Same."
8. Eidinow, "Divine and Human Narratives."
9. Konsolaki-Yannopoulou, "Mycenaean Terracottas," 51.
10. Bellos and Sawidis, "Chemical Pollution."
11. θεὸς ἄμ πόνον ἀνδρῶν, 16.726.
12. König, *Folds of Olympus*, 32.
13. Karavites, *Promise-Giving*, 4.
14. Σπερχειοῖο διιπετέος ποταμοῖο. See Mentzafou, Markogianni. Papadopoulos, Pavlidou, Tziavos, and Dimitriou, "Impacts."
15. Holmes, "Situating Scamander," 44–46; see Hesiod's statement that rivers

were charged by Zeus with raising men (ἀνδρὰς κουρίζουσι, Hesiod, *Theogony*, 347).

16. Hesiod, *Works and Days*, 758–60.
17. Manariotis, "Adverse Effects."
18. Διῒ φίλος or διΐφιλος: Achilles, 1.86; heralds, 8.517; Phoenix, 9.168.
19. E.g., Hecamede, ἔϊκυῖα θεῇσιν, 11.638.
20. αἰθέρος ἐκ δίης, εἰς ἅλα δῖαν, χθὼν δῖα at 16.365, 1.141, 14.347.
21. Apollodorus, *Bibliotheca*, 3.6.8; Pausanias, *Description of Greece*, 8.25.5, 8.25.7.
22. Ready, "Zeus."
23. Διὶ μῆτιν ἀτάλαντε 7.47, 11.200; θοῷ ἀτάλαντος Ἄρηϊ, 8.215–16, 13.528; θεράποντες Ἄρηος, 2.110, 6.67); ὄζος Ἄρηος, 2.540, 12.88; ἀρηίφιλος, e.g., 3.21; ἀρήιος, ἄρειος, 4.407, 15.736; μένος Ἄρηος, 18.264.
24. δῦ δέ μιν Ἄρης.
25. Βροτολοιγός.
26. ὃς θεὸς ἔσκε μετ' ἀνδράσιν.
27. ἐξ ἀνδρῶν χαλέπ' ἄλγε'.
28. Apollodorus, *Bibliotheca* 1.53; Diodorus Siculus, *Diodorus of Sicily*, 4.85.1; Quintus of Smyrna, *Posthomerica*, 1.516.
29. διὰ μῆτιν Ἀθήνης.
30. χαλεποὶ δὲ θεοὶ φαίνεσθαι ἐναργεῖς.
31. αὐτὸς γὰρ ἄκουσα θεοῦ καὶ ἐσέδρακον ἄντην.
32. κελαινεφής, 1.397, 15.46, 21.520; νεφεληγερέτα, 1.511; ἀργικέραυνος, 19.121; τερπικέραυνος, 1.419; ἐρίγδουπος, 5.672; ἐριβρεμέτης, 13.624; ὑψιβρεμέτης, 1.354, 12.68; στεροπηγερέτα, 16.298.
33. ἐγὼν ἡγήσομαι; εἴκελον ἀστεροπῇ.
34. On the similes portraying activities related to shipbuilding see Rood, "Craft Similes."
35. Rudd, *Greenery*, 197.
36. Haubold, *Greece and Mesopotamia*, 64.
37. Janko, "Summary," 554; see also Bintliff, "Regional Geology," 546–47.
38. ψεῦδός κεν φαῖμεν καὶ νοσφιζοίμεθα μᾶλλον.
39. Τρῶες δ' ἐρρίγησαν ὅπως ἴδον αἰόλον ὄφιν / κείμενον ἐν μέσσοισι Διὸς τέρας αἰγιόχοιο.
40. See Clay, *Homer's Trojan Theater*, 57, where this moment is described as in a "magnificent scene of verbal annihilation" for the authorial voice of the *Iliad*.
41. See the scholia cited and discussed in Fenno, "Mist," and Führer, "Diipetés."
42. Apollodorus, *Bibliotecha*, 2.5.8; Diodorus, *Diodorus of Sicily*, 4.32, 49; Hyginus, *Fabulae*, 89.
43. Goldwyn and Kokkini, *John Tzetzes*, xiii.
44. Goldwyn and Kokkini, *John Tzetzes*, xiv.
45. Edwards, *Iliad*, on lines 17.268–69.
46. Eustathius on 17.268–73; see van der Valk, *Commentarii*, vol. 4. See also Kakridis, *Homer Revisited*, 91.

47. Fenno, "Mist," 7. On Homeric "double motivation," see esp. Janko, *Iliad*, 3–4.
48. Fenno, "Mist," 1.
49. Descola, *Beyond Nature*; Almqvist, *Chaos*, esp. 13.
50. Although most of the essays in Iribarren and Koning, *Hesiod*, focus primarily on Hesiod's adumbration of early Greek science and philosophy, the discussions are highly suggestive for the tangled relationship between "nature" and "divinity" in Homeric epics, too.
51. ἔρις πολέμοιο δέδηεν, 17.252.
52. ἐρεμνῇ λαίλαπι ἶσος, 20.51.
53. ἠΰτε πῦρ ἀΐδηλον ἐπιφλέγει ἄσπετον ὕλην/οὔρεος ἐν κορυφῇς, 2.455–56; ὡς εἴ τε πυρὶ χθὼν πᾶσα νέμοιτο, 2.780; ἔπεα νιφάδεσσιν ἐοικότα χειμερίῃσιν, 3.222.
54. κελαινῇ λαίλαπι ἶσος.
55. φλογὶ εἴκελον ἀλκὴν.
56. φλογὶ εἴκελον Ἕκτορα δῖον.
57. Dimock, *Other Continents*, 73–74; see also Holmes, "Situating Scamander," 29–30.
58. Holmes, "Situating Scamander."
59. ἐκλύσθη δὲ θάλασσα ποτὶ κλισίας τε νέας τε / Ἀργείων.
60. König, *Folds of Olympus*, 37.
61. ξὺν Βορέῃ ἀνέμῳ πεπιθοῦσα.
62. Plutarch, *Amatorius*, sec. 20.
63. Lonsdale, *Creatures of Speech*, 10.
64. μητέρα θηρῶν.
65. οἰωνοὶ δὲ περὶ πλέες ἠὲ γυναῖκες.
66. Ogilvy, "Animals," 51.
67. Lonsdale, *Creatures of Speech*, 39.
68. Most critics suppose that the snake has eaten herbs poisonous to man of its own volition in order to replenish its stock of toxic venom (see van der Mije, "Bad Herbs"), but deliberate poisoning of snakes has taken place historically all over the world.
69. On the *Iliad*'s depiction of nature in the similes as menacing and brutal, Bonnafé, *Poésie*, i.22–38; Redfield, *Nature*, 189–92.
70. Clarke, "Between Lions"; Gottschall, "Homer's Human Animal," 280–81.
71. On the pathos of this simile see Rabel, "Agamemnon's *Aristeia*." On the Cebriones simile, see Braund, "Crowing."
72. König, *Folds of Olympus*, 35.

Chapter Five. Loggers

1. Ἴδης ἐν κνημοῖσι πολυπτύχου ὑληέσσης.
2. πολυπῖδαξ.
3. See Efe, Sönmez, Cürebal, and Soykan, "Subalpine Ecosystem"; Knees and Gardner, "Abies nordmanniana."

4. δι' ἠέρος αἰθέρ' ἵκανεν.

5. Uysal, "Overview"; Rix, "Wild about Ida."

6. Öztürk, Uysal, Karabacak, and Çelik, "Plant Species."

7. König, *Folds of Olympus*, 33–34.

8. 11.372, παλαιοῦ δημογέροντος, see also 24.349 and Mackie, "Zeus and Mount Ida," 2.

9. φηγῷ ἐφ' ὑψηλῇ.

10. περικαλλέϊ φηγῷ.

11. Williams, *Deforesting*, 77.

12. εἰνοσίφυλλον.

13. Scherjon, Bakels, MacDonald, and Roebroeks, "Burning the Land."

14. 2 Chronicles 2.

15. McNeill, *Mountains*, 72; Meiggs, *Trees and Timber.*

16. Perlin, *Forest Journey*, 59.

17. Perlin, *Forest Journey*, 58–61.

18. Perlin, *Forest Journey*, 59, 61.

19. Perlin, *Forest Journey*, 63.

20. Hughes, "Theophrastus"; Rubner, "Greek Thought."

21. ἐννῆμαρ μὲν τοί γε ἀγίνεον ἄσπετον ὕλην.

22. Empedocles, *On Nature*, fr. 16.2, in Diels and Kranz, *Die Fragmente der Vorsokratiker*, 1:315.

23. United Nations, *State of the World's Forests.*

24. Perlin, *Forest Journey*, 15–16.

25. Niebauer, "Endangered Amazon Rain Forest," 108, 128.

26. αἰεὶ δὲ πυραὶ νεκύων καίοντο θαμειαί.

27. χίλι' ἄρ' ἐν πεδίῳ πυρὰ καίετο.

28. Kirk, *Iliad*, vol. 2, 341. See also *Iliad*, 1.418.

29. Clay, *Homer's Trojan Theater*, 7–8; König, *Folds of Olympus*, 43.

30. ξύλα κάγκανα.

31. Garcia, "Roasting a 1,000-Pound Cow."

32. ἑλισσομένη περὶ καπνῷ.

33. ζείδωρος ἄρουρα.

34. νῶμα δὲ ξυστὸν μέγα ναύμαχον ἐν παλάμῃσι / κολλητὸν βλήτροισι δυωκαιεικοσίπηχυ.

35. The best discussion remains Meiggs, *Trees and Timber*, 116–53. On the choice of timbers for different parts of ancient ships, see Morrison, Coates, and Rankov, *Athenian Trireme*, 179–81, 205–6, 233–34. There is a great deal of relevant information on the value of timber and its consequent re-use by shipbuilders in Rankov, "For Show, Not Use?"

36. Janko, *Iliad*, 97, 273. See also Rood, "Craft Similes."

37. Kirk, *Iliad*, vol. 1, 131 at line 2.135.

38. The discussion of the tree and timber similes in Meiggs, *Trees and Timber*, 106–8 remains unrivalled and a model of its kind.

39. Schein, *Mortal Hero*, 74.
40. Minchin, *Homer and the Resources of Memory*, 147.
41. δένδρεον ὑψιπέτηλον.
42. This is strangely not addressed as significant in Rood, "Craft Similes," the most detailed discussion of the timber similes.
43. On the colonial mindset of the *Odyssey*, see Hall, *Return of Ulysses*, 75–101, and, on the *Iliad*, Hall, *Introducing*, 49–56.
44. ἐν ἀξύλῳ ἐμπέσῃ ὕλῃ.
45. ἀπὸ δρυὸς οὐδ᾽ ἀπὸ πέτρης.
46. Hesiod, *Works and Days*, 145. On Minoan and Mycenaean divination practices, see Goodison, "Why All This"; for the idea that members of the human race are descended from trees, see Benner, *Selections*, 336. Abram, *Evergreen Ash*, 8–9 points to a similar tradition in Old Norse mythology.
47. Rudd, *Greenery*, 50.

Chapter Six. Farmers

1. *Inque dies magis in montem succedere silvas / cogebant infraque locum concedere cultis*, Lucretius, *Nature of Things*, 5.1370–71.
2. Williams, *Deforesting*, 81.
3. Columella, *On Agriculture*, 2.2.8, 11–12; Hughes and Thirgood, "Deforestation," 64.
4. ἀρούρης καρπὸν.
5. ἐριβώλακα.
6. Hesiod, *Works and Days*, 577–79.
7. Hesiod, *Works and Days*, 31–32.
8. Hesiod, *Works and Days*, 299–302.
9. ἔργον ἐπ᾽ ἔργῳ ἐργάζεσθαι; Hesiod, *Works and Days*, 382–404.
10. Hesiod, *Works and Days*, 456. Perlin, *Forest Journey*, 81–82 adduces this passage when he writes that we ought to call Hesiod's epoch not the Age of Iron but the Age of Wood. See also the excellent remarks in Meiggs, *Trees and Timber*, 260–61.
11. Hesiod, *Works and Days*, 420–47.
12. For "miserable hamlet," see Hesiod, *Works and Days*, 639.
13. Hesiod, *Works and Days*, 505–11; νήριτος ὕλη (511); νήριτος= νήριθμος, countless.
14. Hesiod, *Works and Days*, 553–57.
15. χλαινάων τ᾽ ἀνεμοσκεπέων.
16. Hesiod, *Works and Days*, 159–66, 651–53. See the dazzling discussion of this passage in Clay, *Hesiod's Cosmos*, 177–80.
17. Richardson, "Contest," 2.
18. Rosen, "Aristophanes' *Frogs*," 297–314; Hall, *Theatrical Cast*, 344–49.
19. διὰ λογιστικοῦ προβλήματος. *Contest of Homer and Hesiod*, in Alcidamas, *Mouseion*, 8, 10.

20. For "golden" verses, see West, *Homeric Hymns*, 332–35.

21. πλῆθος ἄπιστον. West, *Homeric Hymns*, 335 n.13.

22. θαυμάσαντες δὲ καὶ ἐν τούτωι τὸν Ὅμηρον οἱ Ἕλληνες ἐπήινουν, ὡς παρὰ τὸ προσῆκον γεγονότων τῶν ἐπῶν. In West, *Homeric Hymns*, 338–41, referring to *Contest of Homer and Hesiod*, 12, 13.

23. Gellner, *Plough*, 17; Morris, *Foragers*, 14.

24. Williams, *Deforesting the Earth*, 74. On the huge impact made in the eastern Mediterranean world by advances in iron technology at the end of the second millennium BCE, see Veldhuijzen, "Rusty Bits."

25. ἀμπελόεσσαν. The vineyard scene on Achilles's shield (18.561–72) is discussed separately later.

26. πολυστάφυλον.

27. See Trapp, "Ajax," 271–72.

28. Arbuckle, "Rise of Cattle Cultures," 290; see Halstead and Isaakidou, "Revolutionary Secondary Products"; Sherratt, "Secondary Exploitation."

29. Arbuckle, "Rise of Cattle Cultures," 290.

30. See especially McInerney, *Cattle*, 74–96 on "epic consumption" of cattle.

31. Konsolaki-Yannopoulou, "Mycenaean Terracottas," 49; similar figurines have been found at Olympia dating to the tenth century BCE.

32. On how much pasturage each sheep would need, see McCune, "How Much Space."

33. Virgil, *Georgics*, 2.520; Theophrastus, *De Causis Plantarum*, 4.8.12.

34. Hughes and Thirgood, "Deforestation," 64.

35. Theophrastus, *De Causis Plantarum*, 5.17.6.

36. Cato and Varro, *On Agriculture*, book 2, introduction, para. 4.

37. Eupolis *Aigides*, fr. 13, in Kassel and Austin, *Poetae Comici Graeci*. This fragment survived because it was quoted in Plutarch, *Table-Talk* 662d; see Storey, *Fragments*.

38. Virgil, *Georgics*, 3.314–15, 2.374–77.

39. Hughes and Thirgood, "Deforestation," 65.

40. Virgil, *Aeneid*, 10.405–9.

41. Cato and Varro, *On Agriculture*, book 2, introduction, para. 4.

42. Apollodorus, *Epitome*, 3.32, in Apollodorus, *Library*, vol. 2.

43. Everest and Özcan, "Determination."

44. Hesiod, *Works and Days*, 606–7.

45. Eubulus, fr. 120, in Kassel and Austin, *Poetae Comici Graeci*.

46. Berdowski, "Heroes and Fish," with further bibliography.

47. Plato, *Ion*, 538d.

48. On this passage and its thematic links with the remainder of book 21, see Kitts, "Wide Bosom."

49. κύν' Ὠρίωνος.

50. Redfield, *Nature*, 193–99; Holmes, "Situating Scamander," 33.

51. Faust, "Die Künstlerische Verwendung," 20–22.

52. Scott, "Dogs in Homer."
53. Euripides, *Bacchae*, 337–40.
54. τραπεζῆες κύνες.
55. Pagliaro, "Proemio," 31–33.
56. Scott, "Dogs in Homer," 227.
57. μέλπηθρον.
58. κύρμα.
59. See Graver, "Dog-Helen," 44.
60. κυνάμυια.
61. κακαὶ κύνες. The gender of the dogs varies in the manuscripts, but the term is pejorative even if we read the masculine *kakoi.*

Chapter Seven. Smiths

Epigraph: Homer, *Iliad*, 17.516–24.

1. Burkert, "Greek Tragedy."
2. Combellack, "Homeric Metaphor"; Fenik, *Typical Battle Scenes*, 114; Ready, "Toil and Trouble."
3. οὐδ᾽ ἔρρηξεν χαλκός.
4. ἀπὸ γὰρ μένος εἵλετο χαλκός (3.294); πάγη δ᾽ ἐν πνεύμονι χαλκός (4.528), see also 20.486; διὰ δ᾽ ἔντερα χαλκὸς ἄφυσσε / δηώσας (14.516–18, 17.345–46); σμερδαλέον κονάβησε· πάλιν δ᾽ ἀπὸ χαλκὸς ὄρουσε / βλημένου, οὐδ᾽ ἐπέρησε (21.593–94).
5. See, e.g., 11.574, 15.137 and 21.69–70, with Stanford, *Greek Metaphor*, 138–39; Moulton, "Homeric Metaphor," 288–89.
6. Ἀρήϊα τεύχεα (16.284); αὐτὸς γὰρ ἐφέλκεται ἄνδρα σίδηρος (16.294 and 19.13). The latter Homeric phrase was sufficiently famous to be quoted, with an obscene twist, by Juvenal (*Satires*, 9.37). In a memorable repurposing, Latin professor Robert Mitchell Henry alluded to it in his tragic history of the failure of the Irish Uprising of Easter 1916. The younger and rasher members of the Irish Volunteers, advocates of immediate armed rebellion, were affected, Henry observes, by mere physical contact with rifles. See Hall, "Sinn Féin," 200.
7. Wertime, "Beginnings," 876.
8. ξίφος ἀργυρόηλον; see Kirk, *Iliad*, vol. 1, 118 on this line (2.45). For a detailed discussion of the extent to which the arms and armor referred to in the *Iliad* correspond with the archaeological evidence, see Gray, "Metal-Working."
9. πλάγχθη δ᾽ ἀπὸ χαλκόφι χαλκός; see Hainsworth, *Iliad*, 265 on line 11.350.
10. Morgan, "Origins," 22. See also Rose, "Homer's *Iliad*," 99.
11. On the evocation of enormous quantities in this list of metal objects, see Sammons, "Gift," especially 272–73.
12. On real-world Bronze Age copper smelting for the production of bronze in Greece, the Aegean, and Cyprus, see especially O'Brien, *Prehistoric Copper Mining*, 55–66.

13. Headrick, *Humans*, 99.
14. Headrick, *Humans*, 100.
15. Cavanagh, Ben-Yosef, and Langgut, "Fuel Exploitation."
16. Book of Psalms 120:4.
17. Cavanagh, Ben-Yosef, and Langgut, "Fuel Exploitation."
18. πολιόν τε σίδηρον.
19. σόλον αὐτοχόωνον.
20. Hainsworth, *Iliad*, 110 on line 9.365.
21. πολύχρυσος πολύχαλκος, πολύχρυσος πολύχαλκος (10.315–16); ἀπερείσι᾽ ἄποινα (10.380).
22. Aeschylus, *Prometheus Victus*, in Aeschylus, *Persians. Seven against Thebes. Suppliants. Prometheus Bound*, 500–503.
23. Healy, *Mining*, 77–78.
24. Pausanias, *Description of Greece*, 10.11.20.
25. Healy, *Mining*, 70, 81.
26. Healy, *Mining*, 81.
27. Cavanagh, Ben-Yosef, and Langgut, "Fuel Exploitation."
28. Cavanagh, Ben-Yosef, and Langgut, "Fuel Exploitation."
29. In general see Craddock, "Conservative Metal Alloying."
30. τηλόθεν ἐξ Ἀλύβης, ὅθεν ἀργύρου ἐστὶ γενέθλη (2.857).
31. Kirk, *Iliad*, vol. 1, 259 on lines 2.856–57.
32. Kirk, *Iliad*, vol. 1, 331 on lines 4.2–3.
33. Kirk, *Iliad*, vol. 1, 141 on lines 2.232–33.
34. For more, see Quinn, *In Search of the Phoenicians*, 49, 70. On the Phoenicians' export trade in the late Bronze and early Iron Ages, especially to Cyprus, see Bell, "Phoenician Trade."
35. Bouzek, "Late Bronze Age," 219 relates Rhesus's gold to sumptuous archaeological finds relating to late Bronze Age gold mines in Rumania.
36. ἀμφὶ δέ οἱ κνημὶς νεοτεύκτου κασσιτέροιο / σμερδαλέον κονάβησε.
37. Polybius, *Histories*, 2.33.3.
38. Lang, "Bronze and Iron," 291.
39. χαλκεύω (18.400); χαλκῆες . . . ἄνδρες (4.187, 4.216, 12.295); χαλκήρης (3.316, etc.); χαλκεοθώρηξ (4.448, 8.62); χαλκοβαρής (15.465, 11.96); χαλκογλώχιν (22.225); εὔχαλκος (20.322); παγχάλκεος (20.102); πολύχαλκον (5.504, 10.315, 18.289); χαλκοκορυστής (5.699, 6.199, 6.398); χαλκοκνήμιδες (7.41); χαλκοπάρηος (12.183, 17.294, 20.397); χαλκόπος (8.41); χαλκότυπος (19.25); χαλκοχίτωνες (1.371, etc.); φράξαντο δὲ νῆας / ἕρκεϊ χαλκείῳ (15.566–67).
40. ἅρματα ποικίλα χαλκῷ.
41. κυνέης διὰ χαλκοπαρήου; χαλκείη κόρυς; αἰχμὴ χαλκείη.
42. πολύς (17.493); νώροπα (2.578); ταναήκεϊ (7.78); φαεινός (12.151); σμερδαλέῳ (13.192); ὀξέϊ (13.338); ἀτειρής (14.25); νηλέϊ (16.761); αἴθοπι (17.3).
43. ταμεσίχροα (4.511); οὐδ᾽ ἀπολήγει / χαλκῷ δηϊόων (17.565–66).
44. κόμπει χαλκὸς ἐπὶ στήθεσσι φαεινὸς (12.151); βόμβησε (13.530); τεύχεα ἀμφα-

ράβησε (21.408); δεινὸν ἔβραχε χαλκός (4.420); μέγα δ᾿ ἔβραχε τεύχεα φωτῶν (16.566).

45. ὁ δ᾿ ἔβραχε χάλκεος Ἄρης (5.859); τόσον ἔβραχ᾿ Ἄρης ἄτος πολέμοιο (5.863).

46. τεύχεα ποικίλ᾿ ἔλαμπε (4.431–32); λάμπετο δουρὸς αἰχμή (6.319); τεύχεσι παμφαίνων ὥς τ᾿ ἠλέκτωρ (6.513); χαλκῷ παμφαῖνον (14.9–11).

47. αἰολομίτρην (5.704–7); ποικίλον ἀστερόεντα (16.133–34); παναίολος (4.186, etc.); αἴθων (4.485, 9.123, 24.233); φαεινός (12.151, 23.561, 4.496, 3.357, 8.272, 13.805, 10.500, 6.219, 18.610); σέλας γένετ᾿ ἠΰτε μήνης (19.375–80). See also Edwards, *Iliad,* 67.

48. Kirk, *Iliad,* vol. 1, 393, on lines 4.510–11.

49. σιδήρειος δ᾿ ὀρυμαγδὸς (17.424); σιδήρεος ἐν φρεσὶ θυμός (22.357); ἐν δὲ πυρὸς μένος ἧκε σιδήρεον (23.177).

50. Kirk, *Iliad,* vol. 2, 139, on lines 5.784–86. This is the only Homeric mention of Stentor. Aristotle (*Politics,* 7.1326b6–7) says that one problem with a city-state that is too large is that nobody can be the town crier "unless he has the lungs of a Stentor."

51. χαλκεοφώνῳ (5.785); χάλκεον ὕπνον (11.241).

52. λάϊνον . . . χιτῶνα. Hainsworth, *Iliad,* 250 on lines 11.241 acknowledges that this locution is very distinctive, but argues, rather, that it means that the sleep of death is unbreakable, "bronze being the toughest material known to the Homeric world"; see also Moulton, "Homeric Metaphor," 284.

53. Mayor, *Gods and Robots,* 132.

54. Mayor, *Gods and Robots,* 7–30.

55. West, *East Face,* 395–96; see also Theognis, lines 77–78, in Gerber, *Tyrtaeus, Solon, Theognis, Mimnermus,* 184–85.

56. οὐδ᾿ εἰ παγχάλκεος εὔχεται εἶναι (20.102); μένος δ᾿ αἴθωνι σιδήρῳ (20.371–72).

Chapter Eight. Achilles's Dystopian Shield

1. Kirk, *Iliad,* vol. 1, 113 on lines 1.586–94; West, *East Face,* 86–87.

2. Gordon, *Before the Bible,* 194–95; West, *East Face,* 388–89.

3. Schironi, *Best of the Grammarians,* 335; see also 274, 276, 292–94.

4. ἄφθιτον ἀστερόεντα μεταπρεπέ᾿ ἀθανάτοισι/χάλκεον.

5. Francis, "Metal Maidens," 8–10; on Hephaestus's automata, see Mayor, *Gods and Robots,* 145–52.

6. ἐν θώματι ἦν ὀρέων τὸ ποιεόμενον; Herodotus, *Histories,* 1.68.1.

7. πρώτη μὲν οὖν ἡ μεταλλευτικὴ καὶ ὑλοτομική: παρασκευαστικαὶ γάρ εἰσιν. Diogenes Laërtius, *Lives and Opinions of Eminent Philosophers,* 3.100.

8. Janko, "Summary," 555.

9. Strabo, *Geography,* 14.5.28.

10. Athenaeus, *Deipnosophistae,* 6.104.

11. Plutarch, *Virtues of Women,* 27.

12. Diodorus Siculus, *Diodorus of Sicily,* 5.38.

13. See Diodorus, *Historical Library*; Hall and Stead, *People's History*, 462.

14. Crielaard, "Homer," 214 with figs.

15. Crielaard, "Homer," 215–16.

16. Crielaard, "Homer," 217–18.

17. Crielaard, "Homer," 215.

18. Markoe, *Phoenician Bronze*, 165, 168.

19. Markoe, *Phoenician Bronze*, 162.

20. Markoe, *Phoenician Bronze*, 169, 176, 195, 171, 181.

21. Markoe, *Phoenician Bronze*, 177, 191, 185.

22. Markoe, *Phoenician Bronze*, 192.

23. CVA Boston 1, pl. 16; 17, 3–4, ca. 560–550 BCE, Museum of Fine Arts, Boston, https://collections.mfa.org/objects/153416.

24. Aeschylus, *Iliad*, 18.35–147, 19.1–39.

25. Kossatz-Deissmann, "Achilleus," 123–27; entry for Achilles in *Lexicon Iconographicum Mythologiae*, 510–25.

26. In the later years of the Trojan war, the son of Telephus, Priam's nephew Eurypylus, led the Mysians to the aid of the Trojans; he was slain by Achilles's son Neoptolemus (Pyrrhus) at the head of the Myrmidons. See Philostratus, *Imagines*, 1.7.

27. Sophocles, *Philoctetes*, 354–90.

28. Homer, *Odyssey*, 11.542–64.

29. Martyniuk, "Playing with Europe," 195–96.

30. Schironi, *Best of the Grammarians*, 524.

31. Rutherford, *Homer*, 25.

32. Leopold, *Round River*, 170.

33. Norrman, *Samuel Butler*, 276.

34. Norrman, *Samuel Butler*, 276.

35. Schironi, *Best of the Grammarians*, 26.

36. Hardie, "*Imago Mundi*," 15. Philia and Neikos can be translated as Love and Strife.

37. Porter, "Hermeneutic Lines," 86.

38. Porter, "Hermeneutic Lines," 87.

39. Geminus, *Eisagoge* 6.12, *sphairikos logos*, translated in Evans and Lennart Berggren, *Geminos's Introduction*, 163; Porter, "Hermeneutic Lines," 88; cf. Heraclitus, *Homeric Questions*, 43, translated in Russell and Konstan, *Heraclitus*.

40. Porter, "Hermeneutic Lines," 91.

41. Porter, "Hermeneutic Lines," 91 n.66; Pfeiffer, *History*, 240.

42. Scholion on Aratus's *Phaenomena*, 26, in Maass, *Commentariorum*, 343, line 17; Porter, "Hermeneutic Lines," 92.

43. Mette, *Sphairopoiia*, xiii and 34.

44. Hardie, "*Imago Mundi*," 13–14; Aeschylus, *Seven against Thebes*, 387–90.

45. Hardie, "*Imago Mundi*" and Hardie, *Cosmos*, 24–31.

46. Hofmann, "Shield of Aeneas."

47. For the pseudo-Hesiodic *Aspis*, "Shield," see Hesiod, *Shield*, 141–320; for Quintus of Smyrna's shields of both Achilles and Eurypylus, see Quintus of Smyrna, *Posthomerica* 5.6–101, 6.198–293; and for Nonnus's shield for Dionysus see Nonnus, *Dionysiaca* 25.387–567.

48. Statius, *Thebaid*, 9.332–38; Sidonius Apollinaris, *Carmina*, 15.17–3.

49. Iribarren, "Shield." For Simonides's apothegm, see Plutarch, *Glory of Athens*, 3.346–47.

50. For Pope's sketch, see BL Add MS 4808, 1712–1724, vol. 2 (ff. 233), British Library, London, https://plasticekphrastic.com/2020/08/10/the-shield-of-achelles/; see also Hall and Stead, *People's History*, 142 n.29.

51. Hubbard, "Nature and Art," 18; Heffernan, *Museum*, 19; see Becker, *Shield*; de Jong, "Shield"; Thein, *Ecphrastic Shields*, esp. the "Introduction" and ch. 2.

52. Heffernan, *Museum*, 22.

53. Auden, "Shield," also published as the title poem of Auden, *Shield of Achilles*. There is an audio recording of Auden himself reading the poem at https://www.youtube.com/watch?v=hpblaBb93fo.

54. Summers, "Or One Could Weep."

55. Heaney, *District*, 71.

56. See Washizuka, "Auden," 10–11.

57. Edwards, *Iliad*, 208.

58. Taplin, "Shield of Achilles," 12, 15, 4; Scully, "Reading," 30.

59. Scully, "Reading," 41.

60. Scully, "Reading," 43.

61. Turner, *Dramas*, 242.

62. Scully, "Reading," 47.

63. σέβας μ' ἔχει εἰσορόωντα; Homer, *Odyssey*, 4.71–75.

64. Nagy, "Just to Look," §15.

65. The citation in the *Life* reads ὀφθέντος δ' αὐτῷ τοῦ Ἀχιλλέως τυφλωθῆναι τὸν Ὅμηρον ὑπὸ τῆς τῶν ὅπλων αὐγῆς. On Hermias's citation, see Cropper, "Scholion." On this tradition, see the insightful comments of Clay, *Homer's Trojan Theater*, 11–12. For the *Life of Homer* commentary, see Pseudo-Herodotus, *Life of Homer*, 6.46–50, https://livingpoets.dur.ac.uk/w/index.php/Pseudo-Herodotus,_Life_of_Homer.

66. See Laird, "Politian's *Ambra*."

67. Poliziano, *Silvae*, 279–84. Ipse ardens clypeo ostentat terramque, fretumque,/ Atque indefessum solem, solisque sororem/Iam plenam, et tacito volventia sidera mundo./Ergo his defixus vates, dum singula visu/Explorat miser incauto, dum lumina figit/Lumina nox pepulit.

68. Poliziano's lines closely imitate those in Calimachus's account of the blinding of Tiresias in his "Bath of Pallas" (Callimachus, *Hymn* 5.68–83), for which see del Lungo, *Poliziano*, 88–89.

69. Edwards, *Iliad*, 208.

70. See Willcock, *Iliad*, 271 on lines 18.509–13.

71. τυκτὸν κακόν.
72. Rutherford, *Homer*, 205: in the *Iliad*'s main narrative, "Ares and Athena are generally at odds (esp. in books 5 and 21), so that this is another point of contrast between the shield and the main body of the poem."
73. Rutherford, *Homer*, 208.
74. Rutherford, *Homer*, 210.
75. Solmsen, "Ilias." The three figures also appear in the Hesiodic *Shield of Heracles*, 156–59, in Hesiod, *Shield*.
76. Wallace, "Cultural Process," 59.
77. Hubbard, "Nature and Art," 22.
78. Willcock, *Iliad*, 270–71 on lines 18.499–500.
79. Mason, "Supreme Court's Bronze Doors," 1396.
80. Mason, "Supreme Court's Bronze Doors," 1396.
81. Taplin, "Shield of Achilles," 6.
82. Philostratus, *Imagines*, 10.15.
83. Philostratus, *Imagines*, 10.17.
84. Wallace, "Cultural Process," 58.
85. Philostratus, *Imagines*, 10.21. See Willcock, *Iliad*, 271 on lines 18.558–60; Rundin, "Politics," 190–91.
86. Willcock, *Iliad*, 71 on lines 18.558–60.
87. Detienne and Svenbro, "Les loups," 220–30.
88. Graver, "Dog-Helen," 45–46.
89. Gellner, *Plough*, 17.
90. Edwards, *Iliad*, 225.
91. Edwards, *Iliad*, 228.
92. Aristarchus, scholion on *Iliad* 18.597–98, in van Thiel, *Aristarch*, p. 211.
93. A photograph and further details can be seen at the website for the Classical Art Research Centre, Oxford: https://www.carc.ox.ac.uk/carc/resources/Introduction-to-Greek-Pottery/Keypieces/blackfigure/francois.
94. See, e.g., Evans, "Palace," 58; Gere, *Knossos*, 81.
95. See the photograph reproduced in Gere, *Knossos*, 83.
96. Edwards, *Iliad*, 229.
97. Hedreen, "Bild."
98. Silius Italicus, *Punica*, 2.395–456.
99. *Oceani gentes;* Silius Italicus, *Punica*, 2.396.
100. Silius Italicus, *Punica*, 2.399–405.
101. See Knight and Riedel, *Aldo Leopold*.
102. Dicks, "Leopold," 178.
103. Leopold, *Sand County*, 242.
104. Lovelock, "Quest for Gaia"; see also Lawrence, *Gaia*.
105. In Leopold, *Round River*, 158–65.
106. MacGillivray, "Round River." Another, more accessible version can be found in Malloch and MacGillivray, "Round River Drive."

107. Leopold, *Round River,* 158–65, at 158.
108. Leopold, *Round River,* 158–59.
109. Leopold, *Round River,* 158.
110. Leopold, *Sand County,* 201.
111. In Leopold, *Sand County,* 104–8.
112. On Ocean's metaphysical importance in the *Iliad,* see the provocative reading in Ali, "Oceanos."
113. In Leopold, *Round River,* 166–73, at 170.
114. Dicks, "Leopold," 200.

Chapter Nine. Humans, Rivers, Floods, and Fire

1. Perlin, *Forest Journey,* 78.
2. Kale, Ejder, Hisar, and Mutlu, "Climate Change"; Kocum and Akgul, "Evaluation."
3. Reid, *Oxford Guide,* 13.
4. Holmes, "Situating Scamander," 30.
5. A. Psilovikos, Mpouras, Papathanasiou, Malamataris, T. Psilovikos, and Spiridis, "Impacts."
6. Aristophanes, *Birds,* 700–702.
7. Ὠκεανόν τε θεῶν γένεσιν (*Iliad* 14.201); ὅς περ γένεσις πάντεσσι τέτυκται (14.246). Orphic *Hymn to Ocean* 83.1–2, Ὠκεανὸν καλέω, πατέρ' ἄφθιτον, αἰὲν ἐόντα,/ἀθανάτων τε θεῶν γένεσιν θνητῶν τ' ἀνθρώπων.
8. Holmes, "Situating Scamander," 50.
9. Albanis, Vosniakos, and Nikolaou, "Axios River."
10. ἀνέρι εἰσάμενος.
11. Holmes, "Situating Scamander," 32.
12. ξεστῆς αἰθούσῃσιν ἐνίζανον.
13. μεμυκὼς ἠΰτε ταῦρος; Nagy, *Greek Mythology,* 325.
14. West, *East Face,* 392.
15. Mills, *Hero,* 67.
16. See Whenzou News, "Water on Fire," with its links to Chinese press articles and photographs.
17. See CBC News, "We Have More to Do."

Epilogue

1. Glacken, *Traces on the Rhodian Shore,* 117–18.
2. Hobbes, *Leviathan,* 13.9.76.
3. Sokolon, "The *Iliad,*" 49.
4. Anhalt, *Enraged,* 3.
5. Christensen, "Epic Glory"; Manguel, "Homer in the Gaza Strip."
6. Stone, *Should Trees Have Standing?* I owe this reference to Sara Monoson.

7. Hughes, "Warfare."
8. Nirappil, Duplain, Timsit, and Villegas, "Grasp at Diplomacy."
9. UNICEF, "Barely a Drop."
10. Davies, "Canadian Alamos Gold"; Gottschlich, "Protest."
11. In 2021 the company reported that the two of its subsidiaries directly involved, Alamos Gold Holdings Coöperatief UA, and Alamos Gold Holdings BV, "will file an investment treaty claim against the Republic of Turkey for expropriation and unfair and inequitable treatment, among other things, with respect to their Turkish gold mining project. The claim will be filed under the Netherlands Turkey Bilateral Investment Treaty (the 'Treaty'), and is expected to exceed $1 billion, representing the value of the Company's Turkish assets." Alamos Gold, "Alamos Gold Announces."
12. For this felicitous formulation, I am deeply indebted to the anonymous reviewer of the manuscript for Yale University Press.

Bibliography

Ancient texts not otherwise cited in the Bibliography may be found online at Loeb Classical Library, https://www.loebclassics.com.

Abram, Christopher. *Evergreen Ash: Ecology and Catastrophe in Old Norse Myth and Literature.* Charlottesville: University of Virginia Press, 2019.

Aeschylus. *Persians. Seven against Thebes. Suppliants. Prometheus Bound,* ed. and trans. Alan H. Sommerstein. Loeb Classical Library 145. Cambridge, MA: Harvard University Press, 2009.

Alamos Gold. "Alamos Gold Announces US$1 Billion Investment Treaty Claim against the Republic of Turkey." Press release, April 20, 2021, https://www.alamosgold.com/news-and-events/news/news-details/2021/Alamos-Gold-Announces-US1-Billion-Investment-Treaty-Claim-Against-the-Republic-of-Turkey/default.aspx.

Albanis, Triantafyllos, A., F. Vosniakos, and Kostas Nikolaou. "Axios River Pollution, Part I: The Pesticides." *Journal of Environmental Protection and Ecology* 10 (2009): 32–36.

Ali, Seemee. "Oceanos' Metaphysical Currents in the *Iliad*." *Les études classiques* 87 (2019): 219–45.

Almqvist, Olaf. *Chaos, Cosmos and Creation in Early Greek Theogonies.* London: Bloomsbury, 2022.

Alram-Stern, Eva, Fritz Blakolmer, Sigrid Deger-Jalkotzy, Robert Laffineur, and Jörg Weilhartner, eds. *Metaphysis: Ritual, Myth and Symbolism in the Aegean Bronze Age.* Leuven, BE: Peeters, 2016.

Anhalt, Emily Katz. *Enraged: Why Violent Times Need Ancient Myths.* New Haven, CT: Yale University Press, 2017.

Apollodorus. *The Library.* Vol. 1: *Books 1–3.9,* trans. James G. Frazer. Loeb Classical Library 121. Cambridge, MA: Harvard University Press, 1921.

———. *The Library.* Vol. 2: *Book 3.10-end, Epitome,* trans. James G. Frazer. Loeb Classical Library 122. Cambridge, MA: Harvard University Press, 1921.

Arbuckle, Benjamin S. "The Rise of Cattle Cultures in Bronze Age Anatolia." *Journal of Eastern Mediterranean Archaeology & Heritage Studies* 2 (2014): 277–97.

Aristophanes. *Birds. Lysistrata. Women at the Thesmophoria*, ed. and trans. Jeffrey Henderson. Loeb Classical Library 179. Cambridge, MA: Harvard University Press, 2000.

Aristotle. *Politics*, trans. H. Rackham. Loeb Classical Library 264. Cambridge, MA: Harvard University Press, 1932.

Athenaeus. *The Learned Banqueters*, ed. and trans. S. Douglas Olson. 8 vols. Loeb Classical Library. Cambridge, MA: Harvard University Press, 2007–2012.

Auden, W. H. "The Shield of Achilles." *Poetry* 81, no. 1 (1952): 3–5.

———. *The Shield of Achilles*. New York: Random House, 1955.

Auerbach, Erich. *Scenes from the Drama of European Literature: Six Essays*. Manchester, UK: Manchester University Press, 1959.

Backman, Jussi, and Antonio Cimino, eds. *Biopolitics and Ancient Thought*. Oxford: Oxford University Press, 2022.

Bakhtin, Mikhail. "Response to a Question from the Novy Mir Editorial Staff." In *Speech Genres and Other Late Essays*, vols. 1–7, eds. Caryl Emerson and Michael Holquist, trans. Vern W. McGee. Austin: University of Texas Press, 1986.

Baldry, H. C. "Hesiod's Five Ages." *Journal of the History of Ideas* 17 (1956): 553–54.

Balmer, Josephine. *Chasing Catullus*. Newcastle upon Tyne, UK: Bloodaxe, 2004.

Barker, Pat. *The Silence of the Girls*. London: Doubleday, 2018.

Becker, A. S. *The Shield of Achilles and the Poetics of Ekphrasis*. Lanham, MD: Rowman & Littlefield, 1995.

Bell, Carol. "Phoenician Trade: The First 300 Years." Pp. 91–105 in *Dynamics of Production in the Ancient Near East, 1300–500 BC*, ed. Juan Carlos Moreno Garcia. Oxford: Oxbow, 2016.

Bellos, Dimitrios, and Thomas D. Sawidis. "Chemical Pollution Monitoring of the River Pinios (Thessalia—Greece)." *Journal of Environmental Management* 76 (2005): 282–92.

Benfey, Christopher. "Introduction." In Simone Weil and Rachel Bespaloff, *War and the "Iliad,"* trans. Mary McCarthy. New York: NYRB Classics, 2005.

Benjamin, Walter. *Illuminations*, ed. Hannah Arendt, trans. Harry Zohn. New York: Schocken, 1969.

Benner, Allen Rogers. *Selections from Homer's "Iliad," with an Introduction, Notes, a Short Homeric Grammar, and a Vocabulary*. New York: Irvington, 1903.

Berdowski, Piotr. "Heroes and Fish in Homer." *Palamedes* 3 (2008): 75–91.

Bergren, Ann L. T. *The Etymology and Usage of PEIRAR in Early Greek Poetry: A Study in the Interrelationship of Metrics, Linguistics and Poetics*. Vol. 2. New York: APA, 1975.

Bergthaller, Hannes, and Peter Mortensen, eds. *Framing the Environmental Humanities*. Leiden: Brill, 2018.

Bespaloff, Rachel. *De l'Iliade*. New York: Brentano's, 1943.

Bintliff, John L. "The Regional Geology and Early Settlement of the Helos Plain."

Pp. 527–50 in William D. Taylour and R. Janko, eds., *Ayios Stephanos: Excavations at a Bronze Age and Medieval Settlement in Southern Laconia*. Athens: British School at Athens, 2008.

Bloomfield-Gadêlha, Connie, and Edith Hall, eds. *Time, Tense and Genre in Ancient Greek Literature*. Oxford: Oxford University Press, forthcoming 2025.

Boitani, Piero. *The Shadow of Ulysses: Figures of a Myth*, trans. Anita Weston. 1992; Oxford: Clarendon Press, 1994.

Bonnafé, Annie. *Poésie, nature et sacré*, vol. 1. Lyon: Maison de l'Orient, 1984.

Bonneuil, Christophe, and Jean-Baptiste Fressoz. *The Shock of the Anthropocene: The Earth, History, and Us*, trans. David Fernbach. London: Verso, 2016.

Borsch, Jonas, and Laura Carrara. "Zwischen Natur und Kultur: Erdleben als Gegenstand der Altertumswissenschaften." In *Erdbeben in der Antike*, 1–4, ed. Borsch and Carrara. Tübingen: Mohr-Siebeck, 2016.

Bosak-Schroeder, Clara. *Other Natures: Environmental Encounters with Ancient Greek Ethnography*. Oakland: University of California Press, 2020.

Boscagli, Maurizia. *Stuff Theory: Everyday Objects, Radical Materialism*. London: Bloomsbury, 2014.

Bouzek, J. "Late Bronze Age Greece and the Balkans: A Review of the Present Picture." *Annual of the British School at Athens* 89 (1994): 217–34.

Bragg, Melvyn. "Plato's Atlantis." Podcast episode of BBC Radio series *In Our Time*, September 22, 2022. https://www.bbc.co.uk/programmes/m001c6t3.

Braund, D. "Crowing over Kebriones the Diver." *Hermathena* 199 (2015): 5–32.

Brockliss, William. *Homeric Imagery and the Natural Environment*. Cambridge, MA: Harvard University Press, 2019.

Burkert, Walter. "Greek Tragedy and Sacrificial Ritual." *Greek, Roman and Byzantine Studies* 7 (1966): 87–121.

Callinicos, A. "The 'New Middle Class' and Socialist Politics." *International Socialism* 2 (1983): 82–119.

Carpenter, R. *Discontinuity in Greek Civilisation*. Cambridge: Cambridge University Press, 1966.

Cato and Varro. *On Agriculture*, trans. W. D. Hooper and Harrison Boyd Ash. Loeb Classical Library 283. Cambridge, MA: Harvard University Press, 1934.

Cavanagh, Mark, Erez Ben-Yosef, and Dafna Langgut. "Fuel Exploitation and Environmental Degradation at the Iron Age Copper Industry of the Timna Valley, Southern Israel." *Scientific Reports* 12 (2022): 1–15.

CBC News. "'We Have More to Do': The 1969 Rouge River Fire Remembered." CBC News, October 9, 2019. https://www.cbc.ca/news/canada/windsor/rouge-river-john-hartig-windsor-detroit-1.5314407.

Chakrabarty, Dipesh. "The Climate of History: Four Theses." *Critical Inquiry* 35 (2009): 199–201.

Chew, Sing C. *World Ecological Degradation: Accumulation, Urbanization and Deforestation 3000 B.C.–A.D. 2000*. Walnut Creek, CA: Rowman and Littlefield, 2001.

Chinn, Christopher. "The Ecological Highway: Environmental Ekphrasis in Statius, *Silvae* 4.3." Pp. 113–29 in Christopher Schliephake, ed., *Ecocriticism, Ecology, and the Cultures of Antiquity*. Lanham, MD: Lexington Books, 2017.

Christensen, Joel. "Epic Glory and Modern War: From the Walls of Troy to the Defence of Kyiv." *Neos Kosmos*, April 26, 2022. https://neoskosmos.com/en/2022/04/26/dialogue/opinion/epic-glory-and-modern-war-from-the-walls-of-troy-to-the-defence-of-kyiv/.

Christenson, Allen J. "Who Shall Be a Sustainer? Maize and Human Mediation in the Maya Popol Vuh." Pp. 93–105 in John Parham and Louise Hutchings Westling, eds., *A Global History of Literature and the Environment*. Cambridge: Cambridge University Press, 2016.

Clark, Timothy. *Ecocriticism on the Edge*. London: Bloomsbury, 2015.

Clarke, Hannah. "Queer Classics." *Eidolon*, July 23, 2019. https://eidolon.pub/queer-classics-b84819356f74.

Clarke, Lindsay. *The War at Troy*. London: HarperCollins, 2004.

Clarke, Michael. "Between Lions and Men: Images of the Hero in the *Iliad*." *Greek, Roman, and Byzantine Studies* 36 (1995): 137–60.

Clay, Jenny Strauss. *Hesiod's Cosmos*. Cambridge: Cambridge University Press, 2003.

———. *Homer's Trojan Theater: Space, Vision, and Memory in the "Iliad."* Cambridge: Cambridge University Press, 2011.

Cline, Eric H. *1177 B.C.: The Year Civilization Collapsed*. 2014; Princeton, NJ: Princeton University Press, 2021.

Cohen, Jeffrey Jerome, and Lowell Duckert. "Eleven Principles of the Elements." Pp. 1–26 in Cohen and Duckert, *Elemental Ecocriticism: Thinking with Earth, Air, Water, and Fire*. Minneapolis: University of Minnesota Press, 2015.

———, eds. *Elemental Ecocriticism: Thinking with Earth, Air, Water, and Fire*. Minneapolis: University of Minnesota Press, 2015.

Cole, Juan. "Think Again: 9/11." *Foreign Policy* 156 (2006): 26–32.

Colebrook, Claire. *Death of the Posthuman: Essays on Extinction*. Vol. 1. London: Open Humanities Press, 2014.

Columella. *On Agriculture*, trans. Harrison Boyd Ash, E. S. Forster, and Edward H. Heffner. 3 vols. Loeb Classical Library. Cambridge, MA: Harvard University Press, 1941–55.

Combellack, Frederick M. "A Homeric Metaphor." *American Journal of Philology* 105 (1984): 247–57.

Cook, Elizabeth Heckendorn. "Remaking Eighteenth-Century Ecologies: Arboreal Mobility." Pp. 156–70 in John Parham and Louise Hutchings Westling, eds., *A Global History of Literature and the Environment*. Cambridge: Cambridge University Press, 2016.

Craddock, P. T. "Conservative Metal Alloying Traditions." In *Archaeometallurgy in Central Europe III*, special issue of *Acta Metallurgica Slovaca* 7 (2000): 175–83.

Crielaard, Jan Paul. "Homer, History and Archaeology: Some Remarks on the

Date of the Homeric World." Pp. 201–88 in Crielaard, ed., *Homeric Questions*. Amsterdam: J.C. Gieben, 2000.

Cropper, Elizabeth. "A Scholion by Hermias to Plato's *Phaedrus* and Its Adaptations in Pietro Testa's *Blinding of Homer* and in Politian's *Ambra*." *Journal of the Warburg and Courtauld Institutes* 43 (1980): 262–65.

Crutzen, P. J., and E. F. Stoermer. "The Anthropocene." *IGBP Newsletter* 41 (2000): 12.

Dalley, Stephanie. "The Natural World in Ancient Mesopotamian Literature." Pp. 21–36 in John Parham and Louise Hutchings Westling, eds., *A Global History of Literature and the Environment*. Cambridge: Cambridge University Press, 2016.

Damrosch, David. *The Buried Book: The Loss and Rediscovery of the Great Epic of Gilgamesh*. New York: H. Holt, 2007.

Dan, A. "Mythic Geography, Barbarian Identities: The Pygmies in Thrace." *Ancient Civilizations from Scythia to Siberia* 20 (2014): 39–66.

Daugherty, Gregory N. "The Cohortes Vigilum and the Great Fire of 64 AD." *Classical Journal* 87 (1992): 229–40.

Davies, Catherine. "Canadian Alamos Gold's Mine Project Destroying Environment in Northwest Turkey." *New Cold War*, August 3, 2019. https://newcold war.org/canadian-alamos-golds-mine-project-destroying-environment-in -northwest-turkey/.

de Jong, I. "The Shield of Achilles: From Metalepsis to Mise en Abyme." *Ramus* 40 (2011): 1–14.

del Carmen Llasat, Maria. "Storms and Floods." Pp. 513–40 in Jamie Woodward, ed., *The Physical Geography of the Mediterranean*. Oxford: Oxford University Press, 2009.

del Lungo, I. *Poliziano, Le Selve e la Strega, prolusioni nello studio fiorentino (1482– 1492)*. Florence: G.C. Sansoni, 1925.

Demetrius, *On Style*, trans. W. Rhys Roberts and Doreen Innes. In *Aristotle, Poetics. Longinus: On the Sublime. Demetrius: On Style*, trans. Stephen Halliwell, W. Hamilton Fyfe, Doreen C. Innes, and W. Rhys Roberts, rev. Donald A. Russell. Loeb Classical Library 199. Cambridge, MA: Harvard University Press, 1995.

Desborough, V. R., R. V. Nicholls, and Mervyn Popham. "A Euboean Centaur." *Annual of the British School at Athens* 65 (1970): 21–30.

Descola, Philippe. *Beyond Nature and Culture*. Chicago: University of Chicago Press, 2013.

Detienne, Marcel, and Jesper Svenbro. "Les loups au festin ou la cité impossible." Pp. 214–37 in M. Detienne and J. P. Vernant, eds., *La cuisine du sacrifice*. Paris: Gallimard, 1979.

Devall, B., and G. Sessions. "Deep Ecology." Pp. 200–207 in M. J. Smith, ed., *Thinking through the Environment: A Reader*. London: Routledge, 1999.

Dickinson, O. T. P. K. "Parallels and Contrasts in the Bronze Age of the Peloponnese." *Oxford Journal of Archaeology* 1 (1982): 125–38.

Dicks, Henry. "Leopold and the Ecological Imaginary." *Environmental Philosophy* 11 (2014): 175–210.

Diels, Hermann, and Walther Kranz. *Die Fragmente der Vorsokratiker.* 3 vols. Berlin: Weidmann, 1960.

Dimock, Wai Chee. *Through Other Continents: American Literature across Deep Time.* Princeton, NJ: Princeton University Press, 2008.

Dimopoulou, N. "Metallurgy and Metalworking in the Harbour-Town of Knossos at Poros Katsambas." Pp. 135–41 in V. Kassianidou and G. Papasavvas, eds., *Eastern Mediterranean Metallurgy and Metalwork in the Second Millennium BC: A Conference in Honour of James D. Muhly, Nicosia 10th–11th October.* Oxford: Oxbow Books, 2009.

Dineen, James. "Catastrophic Libyan Dam Collapse Partly Caused by Climate Change." *New Scientist,* September 19, 2023. https://www.newscientist.com /article/2392811-catastrophic-libyan-dam-collapse-partly-caused-by-climate -change/.

Diodorus Siculus. *Diodorus of Sicily, in Ten Volumes,* trans. C. H. Oldfather. Loeb Classical Library. London: Heinemann, 1933.

———. *The Historical Library of Diodorus the Sicilian,* trans. G. Booth. London: W. Taylor, 1721.

———. *Library of History,* vol. 5: *Books 12.41–13,* trans. C. H. Oldfather, Russel M. Geer, Charles L. Sherman, C. Bradford Welles, and Francis R. Walton. 12 vols. Cambridge, MA: Harvard University Press, 1950–67.

Dolan, John. *The War Nerd Iliad: Modern Prose Translation of Homer's "Iliad."* Minneapolis: Feral House, 2017.

Drake, B. L. "The Influence of Climatic Change on the Late Bronze Age Collapse and the Greek Dark Ages." *Journal of Archaeological Science* 39 (2012): 1862–70.

Drummond, William Hamilton. "Battle of Trafalgar, a Heroic Poem." In Drummond, *Battle of Trafalgar.* Belfast: J. Smythe and D. Lyons, 1835.

Dué, Casey. "Learning Lessons from the Trojan War: Briseis and the Theme of Force." *College Literature* 34 (2007): 229–62.

Dunn, N. E. "The Common Man's 'Iliad.'" *Comparative Literature Studies* 21 (1984): 270–81.

Eckerman, Chris. "Ancient Greek Literature and the Environment: A Case Study with Pindar's *Olympian* 7." Pp. 80–92 in John Parham and Louise Hutchings Westling, eds., *A Global History of Literature and the Environment.* Cambridge: Cambridge University Press, 2016.

Edwards, Mark W. *The "Iliad": A Commentary.* Vol. 5: *Books 17–20.* Cambridge: Cambridge University Press, 1991.

Efe, Recep, Süleyman Sönmez, İsa Cürebal, and Abdullah Soykan. "Subalpine Ecosystem and Possible Impact of Climate Change on Vegetation of Kaz Mountain (Mount Ida—NW Turkey)." Ch. 27 in Munir Ozturk, K. R. Hakeem, I. Faridah-Hanum, and Recep Efe, eds., *Climate Change Impacts on High-Altitude Ecosystems.* Edinburgh: Springer, Cham, 2015.

Eidinow, Esther. "Divine and Human Narratives, Time and Being." In Connie Bloomfield-Gadêlha and Edith Hall, eds. Forthcoming, 2025.

Eisenschiml, Otto, and Ralph Newman. *The American Iliad: The Epic Story of the Civil War as Narrated by Eyewitnesses and Contemporaries.* Indianapolis: Bobbs-Merrill, 1947.

Elyot, Amanda. *The Memoirs of Helen of Troy.* London: Crown, 2005.

Euripides. *Bacchae. Iphigenia at Aulis. Rhesus,* ed. and trans. David Kovacs. Loeb Classical Library 495. Cambridge, MA: Harvard University Press, 2003.

———. *Trojan Women. Iphigenia among the Taurians. Ion,* ed. and trans. David Kovacs. Loeb Classical Library 10. Cambridge, MA: Harvard University Press, 1999.

Evans, Arthur J. "The Palace of Knossos." *Annual of the British School at Athens* 8 (1901): 1–124.

Evans, James, and J. Lennart Berggren. *Geminos's Introduction to the Phenomena: A Translation and Study of a Hellenistic Survey of Astronomy.* Princeton, NJ: Princeton University Press, 2007.

Everest, T., and H. Özcan. "Determination of Soil Erosion Risk of Dümrek Basin Downstream with CORINE Methodology." *COMU Journal of Agriculture Faculty* 5 (2017): 39–47.

Farina, Almo. *Ecosemiotic Landscapes.* Cambridge: Cambridge University Press, 2021.

Faust, M. "Die Künstlerische Verwendung von κύων 'Hund' in Den Homerischen Epen." *Glotta* 48 (1970): 8–31.

Fenik, B. *Typical Battle Scenes in the "Iliad"* (Hermes Einzelschriften 21). Wiesbaden: Steiner, 1968.

Fenno, Jonathan. "The Mist Shed by Zeus in 'Iliad' XVII." *Classical Journal* 104 (2008): 1–9.

Fletchman Smith, B. *Mental Slavery: Psychoanalytical Studies of Caribbean People.* London: Rebus, 2000.

Floyd, Edwin D. "*Kleos Aphthiton:* An Indo-European Perspective on Early Greek Poetry." *Glotta* 58 (1980): 133–57.

Foerster, Donald M. "Homer, Milton, and the American Revolt against Epic Poetry, 1812–1860." *Studies in Philology* 53 (1956): 75–100.

Foucault, Michel. *The Birth of Biopolitics,* trans. Graham Burchell. Basingstoke, UK: Palgrave Macmillan, 1988.

Francis, James A. "Metal Maidens, Achilles' Shield, and Pandora: The Beginnings of 'Ekphrasis,'" *American Journal of Philology* 130 (2009): 1–23.

Frost, K. T. "The *Critias* and Minoan Crete." *Journal of Hellenic Studies* 33 (1913): 189–206.

Führer, Rudolf. "Diipetés." P. 299 in *Lexikon des frühgriechischen Epos,* ed. E.-M. Voigt, vol. 2. Göttingen: Vandenhoeck & Ruprecht, 1982.

Garcia, Salomón. "Roasting a 1,000-Pound Cow Is an Art Form." *Vice,* September 11, 2016. https://www.vice.com/en/article/4x54gq/roasting-entire-cows-is-an-art-form.

Garrard, Greg. "Introduction." Pp. 1–8 in Garrard, ed., *The Oxford Handbook of Ecocriticism*. New York: Oxford University Press, 2014.

———, ed. *The Oxford Handbook of Ecocriticism*. New York: Oxford University Press, 2014.

Gates, Henry Louis, Jr. *Figures in Black: Words, Signs, and the "Racial" Self.* New York: Oxford University Press, 1987.

Gee, Henry. *A (Very) Short History of Life on Earth: 4.6 Billion Years in 12 Chapters.* London: Picador, 2022.

Gellner, Ernst. *Plough, Sword and Book.* Oxford: Blackwell, 1988.

Gemmell, David. *Troy: Fall of Kings.* London: Bantam, 2007.

———. *Troy: Lord of the Silver Bow.* London: Bantam, 2005.

———. *Troy: Shield of Thunder.* London: Bantam, 2006.

George, Margaret. *Helen of Troy: A Novel.* London: Pan, 2006.

Gerber, Douglas E., ed. and trans. *Tyrtaeus, Solon, Theognis, Mimnermus. Greek Elegiac Poetry from the Seventh to the Fifth Centuries BC.* Loeb Classical Library 258. Cambridge, MA: Harvard University Press, 1999.

Gere, Cathy. *Knossos and the Prophets of Modernism.* Chicago: University of Chicago Press, 2009.

Ghosh, Amitav. *The Great Derangement: Climate Change and the Unthinkable.* London: Penguin, 2016.

Gifford, Terry. "The Environmental Humanities and the Pastoral Tradition." In Christopher Schliephake, ed., *Ecocriticism, Ecology, and the Cultures of Antiquity.* Lanham, MD: Lexington, 2017.

Glacken, Clarence J. *Traces on the Rhodian Shore: Nature and Culture in Western Thought from Ancient Times to the End of the Eighteenth Century.* Berkeley: University of California Press, 1976.

Gladstone, William Ewart. *Studies on Homer and the Homeric Age.* 3 vols. Oxford: Oxford University Press, 1858.

Gold, B. K. "Simone Weil: Receiving the *Iliad*." Pp. 360–78 in Rosie Wyles and Edith Hall, eds., *Women Classical Scholars: Unsealing the Fountain from the Renaissance to Jacqueline de Romilly.* Oxford: Oxford University Press, 2016.

Goldwyn, Adam J., and Dimitra Kokkini, trans. *John Tzetzes, Allegories of the "Iliad."* Cambridge, MA: Harvard University Press, 2015.

Goodison, Lucy. "'Why All This about Oak or Stone?': Trees and Boulders in Minoan Religion." *Hesperia Supplements* 42 (2009): 51–57.

Gordon, C. H. *Before the Bible: The Common Background of Greek and Hebrew Civilisations.* London: Collins, 1962.

Gottschall, Jonathan. "Homer's Human Animal: Ritual Combat in the *Iliad*." *Philosophy and Literature* 25 (2001): 278–94.

Göttschlich, Jürgen. "Protest gegen Goldmine in der Türkei: Nicht alles Gold glänzt. Die Tageszeitung, August 21, 2019. https://taz.de/Protest-gegen-Goldmine-in-der-Tuerkei/!5616283/.

Graver, Margaret. "Dog-Helen and Homeric Insult." *Classical Antiquity* 14 (1995): 41–61.

Gray, D. H. F. "Metal-Working in Homer." *Journal of Hellenic Studies* 74 (1954): 1–15.

Graziosi, Barbara, and Emily Greenwood, eds. *Homer in the Twentieth Century: Between World Literature and the Western Canon.* Oxford: Oxford University Press, 2007.

Green, Deborah. "The Garden of Eden in the Hebrew Bible." Pp. 52–64 in John Parham and Louise Hutchings Westling, eds., *A Global History of Literature and the Environment.* Cambridge: Cambridge University Press, 2016.

Greenwood, Emily. *Afro-Greeks: Dialogues between Anglophone Caribbean Literature and Classics in the Twentieth Century.* Oxford: Oxford University Press, 2010.

Griffin, Alan H. F. "Ovid's Universal Flood." *Hermathena* 152 (1992): 39–58.

Grote, George. *A History of Greece.* 12 vols. London: John Murray, 1869.

Guzowska, M., R. Becks, and E. Andersson-Strand. "'She Was Weaving a Great Web.' Textiles in Troia." Pp. 107–11 in M. L. Nosch and R. Laffineur, eds., *KOSMOS. Jewellery, Adornment and Textiles in the Aegean Bronze Age.* Leuven-Liège: Peeters, 2012.

Haber, W. "Anthropozän: Folgen für das Verhältnis der Humanität und Ökologie." Pp. 19–38 in W. Haber, M. Held, and M. Vogt, eds., *Die Welt im Anthropozän.* Munich: Oekom, 2016.

———. "Energy, Food, and Land—The Ecological Traps of Humankind." *Environmental Science and Pollution Research* 14 (2007): 359–65.

Habermas, J. *Legitimation Crisis*, trans. Thomas McCarthy. London: Polity, 1976.

Hahnemann, Carolin. "Book of Paper, Book of Stone: An Exploration of Alice Oswald's *Memorial*." *Arion* 22 (2014): 1–32.

Hainsworth, B. *The "Iliad": A Commentary*, vol. 3. Cambridge: Cambridge University Press, 1993.

Hall, Edith. *Aeschylus' Agamemnon, edited with Introduction, Translation and Commentary.* Liverpool: Liverpool University Press, 2024.

———. "Ancient Greek Literature & Western Identity." Pp. 511–33 in Martin Hose and David Schenker, eds., *Wiley-Blackwell Companion to Greek Literature.* Hoboken, NJ: Wiley-Blackwell, 2015.

———. *Aristotle's Way.* London: Penguin, 2018.

———. "Can the *Odyssey* Ever Be Tragic? Historical Perspectives on the Theatrical Realization of Greek Epic." Pp. 499–523 in M. Revermann and P. Wilson, eds., *Performance, Iconography, Reception.* Oxford: Oxford University Press, 2008.

———. "Classical Epic at the London Fairs: Elkanah Settle's *The Siege of Troy, 1707–1734*." Pp. 439–60 in F. Macintosh, J. McConnell, S. Harrison, and C. Kenward, eds., *Epic Performances.* Oxford: Oxford University Press, 2018.

———. "Classics in Our Ancestors' Communities." Pp. 243–61 in S. Hunt and M. Musie, eds., *Forward with Classics.* London: Bloomsbury, 2018.

———. "Classics Invented: Books, Schools, Universities and Society, 1679–1742." In S. Harrison and C. Pelling, ed., *Classical Scholarship and Its History: Festschrift for Christopher Stray.* Berlin: De Gruyter, 2021.

———. "George Smith: Deciphering the Flood." *Classics and Class*, 2016. https://www.classicsandclass.info/product/203/.

———. "Greek Tragedy and the Politics of Subjectivity in Recent Fiction." *Classical Receptions Journal* 1 (2009): 23–42.

———. "The Immortal Forgotten Other Gang: Blind Orion, Lame Hephaestus, and Dwarf Cedalion." In Ellen Adams, ed., *Disability Studies and the Classical Body*. London: Routledge, 2021.

———. *Introducing the Ancient Greeks: From Bronze Age Seafarers to Navigators of the Western Mind*. London: Norton/Bodley Head, 2014.

———. "Iphigenia and Her Mother at Aulis: A Study in the Revival of a Euripidean Classic." Pp. 3–41 in S. Wilmer and J. Dillon, eds., *Rebel Women*. London: Methuen, 2005.

———. "Karl Marx and the Ruins of Trier." *European Review of History* 18, nos. 5–6 (2011): 783–97.

———. "Materialisms Old & New." Pp. 203–17 in M. Mueller and M. Telò, eds., *The Materialities of Greek Tragedy*. London: Bloomsbury, 2018.

———. "Nine Thousand Years Ago: The Erasure of the Navy from Plato's Atlantis Fictions." In Bloomfield-Gadêlha and Hall, eds. Forthcoming, 2025.

———. "One Precise Day c. 547 BCE: Playing with Time in Lucian's *Charon*." In Bloomfield-Gadêlha and Hall, eds. Forthcoming, 2025.

———. "Ozymandias since the Cold War." Lecture given at Classical Wisdom's Symposium "The End of Empires and the Fall of Nations," August 21, 2021. https://www.youtube.com/watch?v=ylOpBjyL3tI.

———. "Playing Ball with Zeus: Reading Ancient Slavery via Dreams." Pp. 204–32 in R. Alston and L. Proffitt, ed., *Reading Ancient Slavery*. London: Duckworth, 2010.

———. *The Return of Ulysses*. London: IB Tauris, 2008.

———. "Sinn Féin and Ulysses: Between Professor Robert Mitchell Henry and James Joyce." Pp. 193–217 in Isabelle Torrance and Donncha O'Rourke, eds., *Classics and Irish Politics, 1916–2016*. Oxford: Oxford University Press, 2020.

———. *The Theatrical Cast of Athens*. Oxford: Oxford University Press, 2006.

———. *Tony Harrison: Poet of Radical Classicism*. London: Bloomsbury, 2021.

———. "Trojan Suffering, Tragic Gods, and Transhistorical Metaphysics." Pp. 16–33 in Sarah Annes Brown and Catherine Silverstone, eds., *Tragedy in Transition*. Malden, MA: Blackwell, 2007.

———. "Why Is Penelope Still Waiting? The Missing Feminist Reappraisal of the *Odyssey* in Cinema, 1963–2007." Pp. 163–85 in K. P. Nikoloutsos, ed., *Ancient Greek Women in Film*. Oxford: Oxford University Press, 2014.

Hall, Edith, and Fiona Macintosh. *Greek Tragedy and the British Theatre, 1660–1914*. Oxford: Oxford University Press, 2005.

Hall, Edith, and Henry Stead. *A People's History of Classics: Class and Greco-Roman Antiquity in Britain and Ireland, 1689–1939*. London: Routledge Taylor and Francis, 2020.

Halstead, P., and V. Isaakidou. "Revolutionary Secondary Products: The Development and Significance of Milking, Animal-Traction, and Wool-Gathering in Later Prehistoric Europe and the Near East." Pp. 1–76 in T. Wilkinson, S. Sherratt, and J. Bennet, eds., *Interweaving Worlds: Systemic Interactions in Eurasia, 7th to 1st Millennia BC*. Oxford: Oxbow, 2011.

Haraway, Donna. *When Species Meet*. Minneapolis: University of Minnesota Press, 2008.

Hardie, P. R. *Cosmos and Imperium*. Oxford: Clarendon Press, 1986.

———. "*Imago Mundi:* Cosmological and Ideological Aspects of the Shield of Achilles." *Journal of Hellenic Studies* 105 (1985): 11–31.

Harrison, Tony. *Collected Film Poetry*. London: Faber, 2007.

———. *Collected Poems*. London: Faber, 2016.

———. *The Gaze of the Gorgon*. Film. 1992. Viewable at Archive of Performances of Greek & Roman Drama, Oxford University.

Haubold, Johannes. *Greece and Mesopotamia: Dialogues in Literature*. Cambridge: Cambridge University Press, 2013.

Hawkins, Terence. *The Rage of Achilles*. Sacramento, CA: Casperian Books, 2009.

Haynes, Natalie. *A Thousand Ships*. London: Mantle, 2019.

Head, John W. *Deep Agroecology and the Homeric Epics: Global Cultural Reforms for a Natural-Systems Agriculture*. London: Routledge, 2020.

Headrick, Daniel R. *Humans versus Nature: A Global Environmental History*. Oxford: Oxford University Press, 2020.

Healy, J. F. *Mining and Metallurgy in the Greek and Roman World*. London: Thames and Hudson, 1978.

Heaney, Seamus. *District and Circle*. London: Faber & Faber, 2006.

Hedreen, Guy. "Bild, Mythos, and Ritual: Choral Dance in Theseus's Cretan Adventure on the François Vase." *Hesperia* 80 (2011): 491–510.

Heffernan, James A. W. *Museum of Words: The Poetics of Ekphrasis from Homer to Ashbery*, ed. Jerome H. Buckley. Chicago: University of Chicago Press, 1993.

Hegel, G. W. F. *Vorlesungen über die Philosophie der Weltgeschichte*, ed. Georg Lasson. Leipzig: F. Meiner, 1923.

Heiden, Bruce. *Homer's Cosmic Fabrication: Choice and Design in the "Iliad."* Oxford: Oxford University Press, 2008.

Heiken, G., and F. McCoy. "Caldera Development during the Minoan Eruption; Thira, Cyclades, Greece." *Journal of Geophysical Research* 89 (1984): 8441–62.

Heilbrun, Carolyn. "What Was Penelope Unweaving?" Pp. 103–11 in Heilbrun, *Hamlet's Mother and Other Women*. 1985; New York: Columbia University Press, 1990.

Heise, Ursula K. "The Hitchhiker's Guide to Ecocriticism." *PMLA* 121 (2006): 503–16.

———. *Sense of Place, Sense of Planet: The Environmental Imagination of the Global*. Oxford: Oxford University Press, 2008.

Hern, Warren M. "Is Human Culture Carcinogenic for Uncontrolled Population Growth and Ecological Destruction?" *Bioscience* 43 (1993): 768–73.

Herodotus, *The Histories*, trans. Aubrey de Sélincourt, introduction by John Marincola. London: Penguin, 2003.

Hesiod. *The Shield. Catalogue of Women. Other Fragments*, ed. and trans. Glenn W. Most. Loeb Classical Library 503. Cambridge, MA: Harvard University Press, 2018.

———. *Theogony. Works and Days. Testimonia*, ed. and trans. Glenn W. Most. Loeb Classical Library 57. Cambridge, MA: Harvard University Press, 2018.

———. *Works and Days*, trans. Alicia E. Stallings. London: Penguin, 2018.

Hobbes, Thomas. *Leviathan; or, The Matter, Forme, & Power of a Common-wealth, Ecclesiasticall and Civill*. London: A. Crooke, 1651.

Hofmann, Heinz. "The Shield of Aeneas in the Hands of Columbus." *Humanistica Lovaniensia* 66 (2007): 145–79.

Holmes, Brooke. "Foreword." Pp. ix–xiii in Christopher Schliephake, ed., *Ecocriticism, Ecology, and the Cultures of Antiquity*. Lanham, MD: Lexington, 2017.

———. "Situating Scamander: 'Natureculture' in the *Iliad*." *Ramus* 44 (2015): 29–51.

Homer. *The Iliad*, trans. Emily Wilson. New York: Norton, 2017.

———. *The Odyssey*, trans. Walter Shewring. Oxford: Oxford University Press, 1980.

Hubbard, Thomas K. "Nature and Art in the Shield of Achilles." *Arion: A Journal of Humanities and the Classics* 2, no. 1 (1992): 16–41.

Hughes, J. Donald. "Artemis: Goddess of Conservation." *Forest & Conservation History* 34 (1990): 191–97.

———. "Theophrastus as Ecologist." *Environmental Review* 4 (1985): 291–307. Reprinted on vol. 3 of R. Sharples, ed., *Theophrastus Studies: Sources for His Life, Writings, Thought, and Influence*, vol. 3. Leiden: Brill, 1988.

———. "Warfare and the Environment." Pp. 150–62 in Hughes, *Environmental Problems of the Greeks and Romans: Ecology in the Ancient Mediterranean*, 2nd ed. Baltimore: Johns Hopkins University Press, 2014.

Hughes, J. Donald, and J. V. Thirgood. "Deforestation, Erosion, and Forest Management in Ancient Greece and Rome." *Journal of Forest History* 26 (1982): 60–75.

Hughes, Michael. *Country: A Novel*. New York: HarperCollins, 2019.

Hunter, Jack. "Gothic Psychology, the Ecological Unconscious and the Re-Enchantment of Nature." *Gothic Nature*, March 26, 2020. https://gothicnaturejournal.com/gothic-psychology-the-ecological-unconscious/.

Hutchins, Richard. "Interspecies Ethics and Collaborative Survival in Lucretius' *De Rerum Natura*." Pp. 91–111 in Christopher Schliephake, ed., *Ecocriticism, Ecology, and the Cultures of Antiquity*. Lanham, MD: Lexington, 2017.

Hyginus. *Fabulae*, trans. Mary Grant. Topos Text. https://topostext.org/work/206.

İhlas News Agency. "Pollution in River Sakarya 'Rising at Alarming Rate.'" *Hur-*

riyet Daily News, April 11, 2019. https://www.hurriyetdailynews.com/pollution
-in-river-sakarya-rising-at-alarming-rate-142588.

Illingworth, J. R. *Divine Immanence: An Essay on the Spiritual Significance of Matter.*
New York: Macmillan, 1898.

Iraq Body Count. "Iraqi Deaths from Violence, 2003–2011." Iraq Body Count,
January 2, 2012. https://www.iraqbodycount.org/analysis/numbers/2011/.

Iribarren, Leopoldo. "The Shield of Achilles (Ilias XVIII, 478–608) and Simonides' Apothegm on Painting and Poetry (T101 Poltera)." *Poetica* 44 (2012):
289–312.

Iribarren, Leopoldo, and Hugo Koning, eds. *Hesiod and the Beginnings of Greek
Philosophy. Mnemosyne* supplements 455. Leiden: Brill, 2022.

Jahns, S. "Untersuchungen über die holozäne Vegetationsgeschichte von Süddal-
matien und Südgriechenland." Ph.D. diss., University of Göttingen, 1992.

Janko, Richard. "Helen of Troy—or of Lacedaemon? The Trojan War and Royal
Succession in the Aegean Bronze Age." Pp. 118–31 in Jonathan J. Price and
Rachel Zelnick-Abramovitz, eds., *Text and Intertext in Greek Epic and Drama:
Essays in Honor of Margalit Finkelberg.* London: Routledge, 2020.

———. "Summary and Historical Conclusions." Pp. 551–604 in William D. Taylour and R. Janko. *Ayios Stephanos: Excavations at a Bronze Age and Medieval
Settlement in Southern Laconia* (British School at Athens suppl. 44), 2008.

———, ed. *The "Iliad": A Commentary*, vol. 4. Cambridge: Cambridge University
Press, 1992.

Jones, David. *In Parenthesis*, with a Note of Introduction by T. S. Eliot. 1937; London: Faber & Faber, 2014.

Juvenal, *Satires*. In *Juvenal and Persius*, ed. and trans. Susanna Morton Braund.
Loeb Classical Library 91. Cambridge, MA: Harvard University Press, 2004.

Kakridis, Johannes Th. *Homer Revisited*. Lund: Gleerup, 1971.

Kale, Semih, Tuba Ejder, Olcay Hisar, and Fatih Mutlu. "Climate Change Impacts
on Streamflow of Karamenderes River (Çanakkale, Turkey)." *Marine Science
and Technology Bulletin* 5 (2016): 1–6.

Kaniewski, D., E. Paulissen, E. Van Campo, H. Weiss, T. Otto, J. Bretschneider,
and K. Van Lerberghe. "Late Second–Early First Millennium BC Abrupt
Climate Changes in Coastal Syria and Their Possible Significance for the
History of the Eastern Mediterranean." *Quaternary Research* 74, no. 2 (2010):
207–15.

Karavites, Peter. *Promise-Giving and Treaty-Making: Homer and the Near East.*
Leiden: Brill, 1992.

Kassel, Rudolf, and Colin Austin. *Poetae Comici Graeci*, vol. 5. Berlin: De Gruyter,
1986.

Keating, Sara. "Marina Carr: 'There's a Whole World of Women's Work That
Isn't Being Seen': The Playwright Comes Full Circle with the Irish Premiere
of *Hecuba* at Project Arts Centre." *Irish Times*, September 7, 2019.

Kerridge, Richard. "Ecocritical Approaches to Literary Form and Genre: Urgency,

Depth, Provisionality, Temporality." Pp. 361–75 in Greg Garrard, ed., *The Oxford Handbook of Ecocriticism*. New York: Oxford University Press, 2014.

Khan, G. A. "Vital Materiality and Non-Human Agency: An Interview with Jane Bennett." Pp. 42–57 in G. Browning, R. Prokhovnik, and M. Dimova-Cookson, eds., *Dialogues with Contemporary Political Theorists*. Basingstoke, UK: Palgrave Macmillan, 2012.

Kidd, Christopher. "Inventing the 'Pygmy': Representing the 'Other,' Presenting the 'Self.'" *History and Anthropology* 20 (2009): 395–418.

Kirk, Geoffrey Stephen. *The "Iliad": A Commentary*. Vol. 1. Cambridge: Cambridge University Press, 1985.

———. *The "Iliad": A Commentary*. Vol. 2. Cambridge: Cambridge University Press, 1990.

Kitts, Margo. "The Wide Bosom of the Sea as a Place of Death—Maternal and Sacrificial Imagery in *Iliad* 21." *Literature and Theology* 14 (2000): 103–24.

Knapp, A. Bernard, and Sturt W. Manning. "Crisis in Context: The End of the Late Bronze Age in the Eastern Mediterranean." *American Journal of Archaeology* 120 (2016): 99–149.

Knees, S., and M. Gardner. "Abies nordmanniana subsp. equi-trojani." IUCN Red List of Threatened Species, May 10, 2011. https://scite.ai/reports/abies -nordmanniana-ssp-equi-trojani-knees-968Qgk.

Knight, Richard L., and Suzanne Riedel. *Aldo Leopold and the Ecological Conscience*. New York: Oxford University Press, 2002.

Kocum, Esra, and Fusun Akgul. "Evaluation of Environmental Degradation in the Karamenderes River in Relation to Anthropogenic Stressors." *Fresenius Environmental Bulletin* 18 (2009): 762–69.

König, Jason. *The Folds of Olympus. Mountains in Ancient Greek and Roman Culture*. Princeton, NJ: Princeton University Press, 2022.

Konsolaki-Yannopoulou, Eleni. "Mycenaean Terracottas from Funerary Contexts in Troezenia." Pp. 157–69 in *RA-PI-NE-U: Studies on the Mycenaean World Offered to Robert Laffineur for His 70th Birthday*, ed. Jan Driessen. Louvain-la-Neuve: Presses universitaires de Louvain, 2016.

Kossatz-Deissmann, Anneliese. "Achilleus." Pp. 17–200 in *Lexicon Iconographicum Mythologiae Classicae*, Vol. 1. Munich: Artemis Verlag, 1981.

Kovacs, George A. "Truth, Justice, and the Spartan Way: Freedom and Democracy in Frank Miller's *300*." Pp. 381–92 in Lorna Hardwick and Stephen Harrison, eds., *Classics in the Modern World: A Democratic Turn?* Oxford: Oxford University Press, 2013.

Krapf, Tobias. "Symbolic Value and Magical Power: Examples of Prehistoric Objects Reused in Later Contexts in Euboea." Pp. 531–34 in Eva Alram-Stern et al., eds., *Metaphysis: Ritual, Myth and Symbolism in the Aegean Bronze Age*. Leuven, BE: Peeters, 2016.

Laffineur, Robert, and Thomas G. Palaima, eds. *Zoia: Animal-Human Interactions in the Aegean Middle and Late Bronze Age* (*Aegaeum* 45). Leuven, BE: Peeters, 2021.

Lahr, John. "Inventing the Enemy." *New Yorker* 18 (1993): 103–6.

Laird, Andrew. "Politian's *Ambra* and Reading Epic Didactically." Pp. 27–47 in Monica Gale, ed., *Latin Epic and Didactic Poetry: Genre, Tradition and Individuality*. Swansea: University Press of Wales, 2004.

Lang, Andrew. "Bronze and Iron in Homer." *Revue Archéologique* 7 (1906): 280–96.

Lanone, Catherine. "L'état d'exception dans *The Silence of the Girls* de Pat Barker." *Études britanniques contemporaines, OpenEdition Journals* 58 (2020): http://journals.openedition.org/ebc/8286.

Lawrence, E. *Gaia: The Growth of an Idea*. New York: St. Martin's, 1990.

Leopold, Aldo. "Land as Circulatory System." Unpublished manuscript, 1943.

———. *Round River*. Oxford: Oxford University Press, 1953.

———. *A Sand County Almanac, and Sketches Here and There*. Oxford: Oxford University Press, 1949.

Levi, Peter. *Atlas of the Greek World*. New York: Facts on File, 1984.

Lexicon Iconographicum Mythologiae Classicae. Vol. 1. Zürich: Artemis & Winkler, 1981.

Lewis, Simon L., and Mark A Maslin. "Defining the Anthropocene." *Nature* 519 (2015): 171–80.

Liritzis, Ioannis, Alexander Westra, and Changhong Miao. "Disaster Geoarchaeology and Natural Cataclysms in World Cultural Evolution: An Overview." *Journal of Coastal Research* 35 (2019): 1307–30.

Lisboa, Maria Manuel. *The End of the World: Apocalypse and Its Aftermath in Western Culture*. Cambridge: Open Book Publishers, 2011.

Lloret, Francisco, Josep Pinol, and Marc Castellnou. "Wildfires." Pp. 541–58 in Jamie Woodward, ed., *The Physical Geography of the Mediterranean*. Oxford: Oxford University Press, 2009.

Longinus, *On the Sublime*, ed. and trans. W. Hamilton Fyfe. In *Aristotle, Poetics. Longinus: On the Sublime. Demetrius: On Style*, trans. Stephen Halliwell, W. Hamilton Fyfe, Doreen C. Innes, and W. Rhys Roberts, rev. Donald A. Russell. Loeb Classical Library 199. Cambridge, MA: Harvard University Press, 1995.

Longley, Michael. *Ghost Orchid*. London: Cape Poetry, 1995.

Lonsdale, Steven H. *Creatures of Speech: Lion, Herding, and Hunting Similes in the "Iliad."* Stuttgart: Teubner, 1990.

Lorimer, H. M. *Homer and the Monuments*. London: Macmillan, 1950.

Lovelock, James. *Gaia: A New Look at Life on Earth*. Oxford: Oxford University Press, 1979.

———. "The Quest for Gaia." *New Scientist* 65 (1979): 304.

Lowe, Celia, et al. *Anthropocene Unseen: A Lexicon*. Galeta, CA: Punctum Books, 2020.

Lucretius. *The Nature of Things*, trans. Alicia E. Stallings. London: Penguin, 2007.

Luján, Eugenio R., and Alberto Bernabé. "Ivory and Horn Production in Mycenaean Texts." Pp. 627–38 in Marie Louise Nosch and Robert Laffineur, eds., *KOSMOS: Jewellery, Adornment, and Textiles in the Aegean Bronze Age*. Leuven-Liège: Peeters, 2012.

Maas, G. S., and M. G. Macklin. "The Impact of Recent Climate Change on Flooding and Sediment Supply within a Mediterranean Catchment, Southwestern Crete, Greece." *Earth Surface Processes* 27 (2002): 1087–105.

Maass, Ernst, ed. *Commentariorum in Aratum reliquiae*. Berlin: Weidmann, 1898.

MacGillivray, James. "Round River." *The Press* (Oscado, MI), August 10, 1906.

Mackie, C. "Zeus and Mount Ida in Homer's *Iliad*." *Antichthon* 48 (2014): 1–13.

Malloch, Douglas, and James MacGillivray. "The Round River Drive." *American Lumberman*, April 25, 1914.

Malouf, David. *Neighbours in a Thicket: Poems*. St. Lucia: University of Queensland Press, 1974.

———. *Ransom*. London: Vintage, 2009.

Manariotis, Ioannis D. "Adverse Effects on Alfeios River Basin and an Integrated Management Framework Based on Sustainability." *Environmental Management* 34 (2004): 261–69.

Manfredi, Valerio Massimo. *The Talisman of Troy*. London: Pan, 2004.

Manguel, Alberto. "Homer in the Gaza Strip." *El País*, November 1, 2023. https://english.elpais.com/opinion/2023-11-01/homer-in-the-gaza-strip.html.

Marinatos, Nanno. "Myth, Ritual, Symbolism and the Solar Goddess in Thera." Pp. 3–10 in Eva Alram-Stern, Fritz Blakolmer, Sigrid Deger-Jalkotzy, Robert Laffineur, and Jörg Weilhartner, eds., *Metaphysis: Ritual, Myth and Symbolism in the Aegean Bronze Age*. Leuven, BE: Peeters, 2016.

Markoe, Glenn. *Phoenician Bronze and Silver Bowls from Cyprus and the Mediterranean*. University of California Publications, Classical Studies, vol. 26. Berkeley: University of California Press, 1985.

Marrou, H.-I. *A History of Education in Antiquity*, trans. George Lamb. New York: Sheed & Ward, 1956.

Martin, Richard. "Introduction" to *The "Iliad" of Homer*, reissue of translation by Richmond Lattimore. Chicago and London: University of Chicago Press, 2011.

———. *The Language of Heroes: Speech and Performance in the "Iliad."* Ithaca, NY: Cornell University Press, 1989.

Martyniuk, Irene. "Playing with Europe: Derek Walcott's Retelling of Homer's *Odyssey*." *Callaloo* 28 (2005): 188–99.

Mason, David. "The Supreme Court's Bronze Doors." *American Bar Association Journal* 63 (1977): 1395–99.

Mathews, Freya. *The Ecological Self*. London: Routledge, 2006.

Mayor, Adrienne. *The First Fossil Hunters: Paleontology in Greek and Roman Times*. Princeton, NJ: Princeton University Press, 2000.

———. *Gods and Robots: Myths, Machines, and Ancient Dreams of Technology*. Princeton, NJ: Princeton University Press, 2018.

McAnany, Patricia, and Norman Yofee. *Questioning Collapse*. Cambridge: Cambridge University Press, 2010.

McConnell, Justine. *Black Odysseys: The Homeric Odyssey in the African Diaspora since 1939*. Oxford: Oxford University Press, 2013.

McCune, Kathy. "How Much Space Do You Need to Raise Sheep." February 23, 2023. https://familyfarmlivestock.com/how-much-space-do-you-need-to -raise-sheep/#open.

McInerney, Jeremy. *The Cattle of the Sun: Cows and Culture in the World of the Ancient Greeks.* Princeton, NJ: Princeton University Press, 2010.

McInnes, N. *The Western Marxists.* London: Alcove Press, 1972.

McNeill, J. R. "The Classics Today." Speech at the Official Launch of the New ANU Bachelor of Classical Studies and the Classics Endowment at the Australian National University in Canberra, September 11, 2009. http://www .anu.edu.au.

———. *The Mountains of the Mediterranean World.* Cambridge: Cambridge University Press, 1992.

Medovoi, Leerom. "The Biopolitical Unconscious: Toward an Eco-Marxist Literary Theory." *Mediations* 24 (2009): 122–38.

Meiggs, Russell. *Trees and Timber in the Ancient Mediterranean World.* Oxford: Clarendon Press, 1982.

Mentzafou, A., V. Markogianni, A. Papadopoulos, A. Pavlidou, C. Tziavos, and E. Dimitriou. "The Impacts of Anthropogenic and Climatic Factors on the Interaction of Spercheios River and Maliakos Gulf, the Aegean Sea." Pp. 1–33 in R. D. Deshpande, ed., *The Handbook of Environmental Chemistry.* Berlin: Springer, 2020.

Mette, H. J. *Sphairopoiia; Untersuchungen zur Kosmologie des Krates von Pergamon.* Munich: Beck, 1936.

Middleton, Guy D. "Nothing Lasts Forever: Environmental Discourses on the Collapse of Past Societies." *Journal of Archaeological Research* 20 (2012): 257–307.

———. *Understanding Collapse: Ancient History and Modern Myths.* Cambridge: Cambridge University Press, 2017.

Mills, Donald H. *The Hero and the Sea: Patterns of Chaos in Ancient Myth.* Wauconda, IL: Bolchazy-Carducci, 2002.

Minchin, E. *Homer and the Resources of Memory: Some Applications of Cognitive Theory to the "Iliad" and the "Odyssey."* Oxford: Oxford University Press, 2001.

Möller, Astrid. "Experience and Expectations: Hesiod on Work, Justice, and Environment." Pp. 25–36 in Christopher Schliephake, ed., *Ecocriticism, Ecology, and the Cultures of Antiquity.* Lanham, MD: Lexington Books, 2017.

Moormann. Eric M. *Pompeii's Ashes: The Reception of the Cities Buried by Vesuvius in Literature, Music, and Drama.* Boston: De Gruyter, 2015.

Morgan, C. "The Origins of Pan-Hellenism." Pp. 18–44 in Nanno Marinatos and Robin Hägg, eds., *Greek Sanctuaries.* London: Routledge, 1993.

Morris, Ian. *Foragers, Farmers and Fossil Fuels: How Human Values Evolve.* Princeton, NJ: Princeton University Press, 2015.

Morrison, J. S., J. F. Coates, and N. B. Rankov. *The Athenian Trireme: The History and Reconstruction of an Ancient Greek Warship.* 2nd ed. Cambridge: Cambridge University Press, 2000.

Most, Glenn W., ed. *Hesiod. The Shield. Catalogue of Women. Other Fragments*. Cambridge, MA: Harvard University Press, 2018.

Moulton, Carroll. "Homeric Metaphor." *Classical Philology* 74 (1979): 279–93.

Mylonas, George. *Mycenae and the Mycenaean Age*. Princeton, NJ: Princeton University Press, 1966.

Naess, A. "Self-Realization: An Ecological Approach to Being in the World." Pp. 225–39 in G. Sessions, ed., *Deep Ecology for the Twenty-First Century*. Boston: Shambhala, 1995.

Nagy, G. *The Best of the Achaeans: Concepts of the Hero in Archaic Greek Poetry*. Baltimore, MD: Johns Hopkins University Press, 1979.

———. *Greek Mythology and Poetics*. Ithaca, NY: Cornell University Press, 1992.

———. "Just to Look at All the Shining Bronze Here, I Thought I'd Died and Gone to Heaven: Seeing Bronze in the Ancient Greek World." *Classical Inquiries*, February 18, 2016. https://classical-inquiries.chs.harvard.edu/just-to-look-at-all-the-shining-bronze-here-i-thought-id-died-and-gone-to-heaven-seeing-bronze-in-the-ancient-greek-world/.

Nardizzi, Vin. *Wooden Os: Shakespeare's Theatres and England's Trees*. Toronto: University of Toronto Press, 2013.

Naveh, Z., and J. Dan. "The Human Degradation of Mediterranean Landscapes in Israel." Pp. 173–90 in F. de Castri and H. A. Mooney, eds., *Mediterranean Type Ecosystems*. New York: Springer, 1973.

Newbold, R. F. "Social and Economic Consequences of the A.D. 64 Fire at Rome." *Latomus* 33 (1974): 858–69.

Niebauer, Kevin. "The Endangered Amazon Rain Forest in the Age of Ecological Crisis." Pp. 107–28 in Frank Uekötter, ed., *Exploring Apocalyptica*. Pittsburgh PA: University of Pittsburgh Press, 2018.

Nirappil, Fenit, Julian Duplain, Annabelle Timsit, and Paulina Villegas. "A Grasp at Diplomacy as Fighting Grinds on in Ukraine." *Washington Post*, May 22, 2022. https://www.washingtonpost.com/national-security/2022/05/22/russia-ukraine-zelensky-poland/.

Nonnus. *Dionysiaca*, trans. W. H. D. Rouse. 3 vols. Loeb Classical Library. Cambridge, MA: Harvard University Press, 1940.

Norrman, Ralf. *Samuel Butler and the Meaning of Chiasmus*. London: St. Martin's, 1986.

O'Brien, William. *Prehistoric Copper Mining in Europe: 5500–500 BC*. Oxford: Oxford University Press, 2015.

O'Connor, James. *Natural Causes: Essays in Ecological Marxism*. New York: Guilford Press, 1998.

Ogilvy, James. "Animals in the *Iliad*." *Echos du monde Classique* 2 (1972): 49–53.

Olsen, Sarah, and Mario Telò, eds. *Queer Euripides: Re-readings in Greek Tragedy*. London: Bloomsbury Academic, 2022.

Oppenheimer, C., and D. M. Pyle. "Volcanoes." Pp. 435–68 in J. C. Woodward, ed., *The Physical Geography of the Mediterranean*. Oxford: Oxford University Press, 2009.

Osborn, Fairfield. *The Limits of the Earth.* Toronto: McClelland and Stewart, 1953.

Oswald, Alice. *Memorial.* London: Faber and Faber, 2011.

Ovid. *Metamorphoses,* trans. Frank Justus Miller, rev. G. P. Goold. 2 vols. Loeb Classical Library. Cambridge, MA: Harvard University Press, 1916.

Öztürk, M., İ Uysal, E. Karabacak, and S. Çelik. "Plant Species Microendemism, Rarity and Conservation of Pseudo-Alpine Zone of Kazdağı (Mt. Ida) National Park—Turkey." *Procedia—Social and Behavioral Sciences* 19 (2011): 778–86.

Pack, Roger Ambrose. *The Greek and Latin Literary Texts from Greco-Roman Egypt.* 2nd ed. Ann Arbor: University of Michigan Press, 1967.

Pagliaro, A. "Il Proemio dell' *Iliade.*" In *Nuovi saggi di critica semantica,* 3–46. Rev. ed. Messina: G. D'Anna, 1971.

Palaima, Thomas G. "The Metaphysical Mind in Mycenaean Times and in Homer." Pp. 479–84 in Eva Alram-Stern, Fritz Blakolmer, Sigrid Deger-Jalkotzy, Robert Laffineur, and Jörg Weilhartner, eds., *Metaphysis: Ritual, Myth and Symbolism in the Aegean Bronze Age.* Leuven, BE: Peeters, 2016.

Papadopoulos, Gerassimos. "Tsunamis." Pp. 493–512 in Jamie Woodward, ed., *The Physical Geography of the Mediterranean.* Oxford: Oxford University Press, 2009.

Parham, John, and Louise Hutchings Westling, eds. *A Global History of Literature and the Environment.* Cambridge: Cambridge University Press, 2016.

———. "Introduction." Pp. 1–17 in Parham and Westling, eds., *A Global History of Literature and the Environment.* Cambridge: Cambridge University Press, 2016.

Parker, Lynne. "Hecuba." RTE, September 30, 2019. https://www.rte.ie/culture/2019/0910/1075010-hecuba-marina-carr-and-lynne-parker-reimagine-greek-legend/.

Pausanias. *Guide to Greece,* trans. Peter Levi. 2 vols. London: Penguin, 1984.

Pavlou, Stel. *Gene.* London: Pocket, 2005.

Payne, Mark. *Flowers of Time: On Postapocalyptic Fiction.* Princeton, NJ: Princeton University Press, 2020.

Pease, Donald. "Fear, Rage, and the Mistrials of Representation in *The Red Badge of Courage.*" Pp. 168–75 in Eric J. Sundquist, ed., *American Realism: New Essays.* Baltimore: Johns Hopkins University Press, 1982.

Perlin, John. *A Forest Journey: The Role of Wood in the Development of Civilization.* Cambridge, MA: Harvard University Press, 1991.

Pfeiffer, Rudolf. *History of Classical Scholarship from the Beginnings to the End of the Hellenistic Age.* Oxford: Clarendon, 1968.

Philostratus. *Imagines.* In *Philostratus the Elder, Imagines. Philostratus the Younger, Imagines. Callistratus, Descriptions,* trans. Arthur Fairbanks. Loeb Classical Library 256. Cambridge, MA: Harvard University Press, 1931.

Pindar. *Nemean Odes. Isthmian Odes. Fragments,* ed. and trans. William H. Race. Loeb Classical Library 485. Cambridge, MA: Harvard University Press, 1997.

Plato. *Ion.* In *Plato. Statesman. Philebus. Ion,* trans. Harold North Fowler and W. R. M. Lamb. Loeb Classical Library 164. Cambridge, MA: Harvard University Press, 1925.

Plutarch. *Amatorius.* In *Moralia,* vol. 9, trans. Edwin L. Minar, F. H. Sandbach, and

W. C. Helmbold. Loeb Classical Library 425. Cambridge, MA: Harvard University Press, 1961.

———. *Virtues of Women*. In *Moralia*, vol. 3, trans. Frank Cole Babbitt. Loeb Classical Library 245. Cambridge, MA: Harvard University Press, 1931.

Poliziano, Angelo. *Silvae*, ed. and trans. Charles Fantazzi. Cambridge, MA: Harvard University Press, 2004.

Polybius. *Histories*, trans. Evelyn S. Shuckburgh. London: Macmillan, 1889.

Porter, James I. "Hermeneutic Lines and Circles: Aristarchus and Crates on the Exegesis of Homer." Pp. 67–114 in Robert Lamberton and John J. Kennedy, eds., *Homer's Ancient Readers*. Princeton, NJ: Princeton University Press, 1992.

PR Newswire. "White House Meets with Hollywood Leaders to Explore Ways to Win War against Terror." *PR Newswire*, November 11, 2001.

Psilovikos, A., G. Mpouras, T. Papathanasiou, D. Malamataris, T. Psilovikos, and A. Spiridis. "Impacts of Wildfires on Surface Runoff and Erosion: The Case Study of a Fire Event in Pelion Area, Greece." 17th International Conference on Environmental Science and Technology, Athens, Greece, September 2021. https://www.academia.edu/94896410/Impacts_of_Wildfires_on_Surface_Runoff_and_Erosion_The_Case_Study_of_a_Fire_Event_in_Pelion_Area_Greece.

Quinn, Josephine Crawley. *In Search of the Phoenicians*. Princeton, NJ: Princeton University Press, 2018.

Quintus Smyrnaeus. *Posthomerica*, ed. and trans. Neil Hopkinson. Loeb Classical Library 19. Cambridge, MA: Harvard University Press, 2018.

Rabel, Robert R. "Agamemnon's *Aristeia: Iliad* 11.101–21." *Syllecta Classica* 2 (1990): 1–7.

Rackham, Oliver. "Ecology and Pseudo-Ecology: The Example of Ancient Greece." Pp. 1–43 in G. Shipley and John Salmon, eds., *Human Landscapes in Classical Antiquity*. London: Routledge, 1996.

Rankov, Boris. "For Show, Not Use? Ptolemy IV Philopator and His Forty." Pp. 33–84 in Rita Amedick, Heide Froning, and Winfried Held, eds., *Marburger Winckelmann-Programm*. Marburg: Martina Klein, 2017.

Rappenglück, Barbara, and Michael Rappenglück. "Does the Myth of Phaethon Reflect an Impact?—Revising the Fall of Phaethon and Considering a Possible Relation to the Chiemgau Impact." Pp. 101–9 in *Ancient Watching of Cosmic Space and Observation of Astronomical Phenomena—Mediterranean Archaeology and Archaeometry, Proceedings of the International Conference on Archaeoastronomy*. European Society for Astronomy in Culture (SEAC), Basel, 2006.

Ready, Jonathan L. "Toil and Trouble: The Acquisition of Spoils in the *Iliad*." *Transactions of the American Philological Association* 137 (2007): 3–43.

———. "Zeus, Ancient Near Eastern Notions of Divine Incomparability, and Similes in the Homeric Epics." *Classical Antiquity* 31 (2012): 56–91.

Redfield, James. *Nature and Culture in the "Iliad": The Tragedy of Hector*. Durham, NC: Duke University Press, 1975.

Reid, Jane Davidson. *The Oxford Guide to Classical Mythology in the Arts*. Vol. 1. New York: Oxford University Press, 1993.

Reynolds, Margaret. "The 'Poem of Force' in Australia: David Malouf, Ransom and Chloe Hooper, *The Tall Man*." Pp. 95–209 in J. McConnell and E. Hall, eds., *Ancient Greek Myth in World Fiction since 1989*. London: Bloomsbury, 2016.

Richardson, Nicholas J. "The Contest of Homer and Hesiod and Alcidamas' *Mouseion*." *Classical Quarterly* 31 (1981): 1–10.

———, ed. *The "Iliad": A Commentary*, vol. 6. Cambridge: Cambridge University Press, 1993.

Rix, M. "Wild about Ida: The Glorious Flora of Kaz Dagi and the Vale of Troy." *Cornucopia* 26 (2002): 54–74.

Roberts, Neil, and Jane Reed. "Lakes, Wetlands and Holocene Environmental Change." Pp. 255–86 in Jamie Woodward, ed., *The Physical Geography of the Mediterranean*. Oxford: Oxford University Press, 2009.

Rood, Naomi. "Craft Similes and the Construction of Heroes in the *Iliad*." *Harvard Studies in Classical Philology* 104 (2008): 19–43.

Rose, Peter W. "Homer's *Iliad*: Alienation from a Changing World." Pp. 93–133 in Rose, *Class in Archaic Greece*. Cambridge: Cambridge University Press, 2012.

———. *Sons of the Gods, Children of Earth*. Ithaca, NY: Cornell University Press, 1992.

Rosen, Ralph. "Aristophanes' *Frogs* and the *Contest of Homer and Hesiod*." *Transactions of the American Philological Association* 134 (2004): 295–322.

Rossetti, William Michael. *Notes on the Royal Academy Exhibition 1868*, pt. 1. London: John Camden Hotten, 1868.

Roszak, Theodore. "Awakening the Ecological Unconscious. Ecopsychology: Healing Our Alienation from the Rest of Creation." *Exploring Our Interconnectedness* 34 (1993): 48.

———. *The Voice of the Earth: An Exploration of Ecopsychology*. 2nd ed. Grand Rapids, MI: Phanes Press, 1992.

Rowland, Susan. *The Ecocritical Psyche: Literature, Evolutionary Complexity and Jung*. London: Routledge, 2012.

Rubner, Heinrich. "Greek Thought and Forest Science." *Environmental Review* 9 (1985): 277–95.

Rudd, Gillian. *Greenery: Ecocritical Readings of Late Medieval English Literature*. Manchester, UK: Manchester University Press, 2007.

Rundin, John. "A Politics of Eating: Feasting in Early Greek Society." *American Journal of Philology* 117 (1996): 179–215.

Ruskin, John. "The Mystery of Life and Its Arts." Pp. 93–137 in R. H. Martley, ed., *The Afternoon Lectures on Literature and Art. Delivered in Theatre of the Royal College of Science, S. Stephen's Green, Dublin, in the Years 1867 & 1868*. Dublin: William McGee, 1869.

Russell, D. A., and David Konstan. *Heraclitus: Homeric Problems*. Atlanta: Society of Biblical Literature, 2005.

Rutherford, R. B., ed. *Homer: "Iliad" Book XVIII.* Cambridge: Cambridge University Press, 2019.

Sackett, L. H., and M. R. Popham. "Lefkandi: A Euboean Town of the Bronze Age and the Early Iron Age (2100–700 B.C.)." *Archaeology* 25 (1972): 8–19.

Sammons, Benjamin. "Gift, List & Story in *Iliad* 9.115–61." *Classical Journal* 103 (2008): 353–79.

Schadewaldt, W. *Iliasstudien.* Leipzig: Sächsische Akademie der Wissenschaften, 1938.

Schaeffer, C. F. A. *Stratigraphie comparée et Chronologie de l'Asie Occidentale (Ille et IIe Millénaires), Syrie, Palestine, Asie Mineure, Chypre, Perse et Caucase.* Oxford: Griffith Institute, Ashmolean Museum, and Oxford University Press, 1948.

Schama, Simon. *Landscape and Memory.* London: HarperCollins, 1995.

Schein, Seth. *The Mortal Hero: An Introduction to Homer's "Iliad."* Berkeley: University of California Press, 1984.

———. "Reading Homer in Dark Times: Rachel Bespaloff's *On the Iliad.*" *Arion* 26 (2018): 17–36.

Scherjon, Fulco Corrie Bakels, Katharine MacDonald, and Wil Roebroeks. "Burning the Land: An Ethnographic Study of Off-Site Fire Use by Current and Historically Documented Foragers and Implications for the Interpretation of Past Fire Practices in the Landscape." *Current Anthropology* 56 (2015): 299–326.

Schironi, Francesca. *The Best of the Grammarians: Aristarchus of Samothrace on the Iliad.* Ann Arbor: Michigan University Press, 2018.

Schliephake, Christopher, ed. *Ecocriticism, Ecology, and the Cultures of Antiquity.* Lanham, MD: Lexington Books, 2017.

———. *The Environmental Humanities and the Ancient World.* Cambridge: Cambridge University Press, 2020.

Schliephake, Christopher, and Evi Zemanek, eds. *Anticipatory Environmental (Hi)Stories from Antiquity to the Anthropocene.* Lanham, MD: Lexington Books, 2023.

———. "Introduction." Pp. 1–21 in Christopher Schliephake and Evi Zemankek, eds. *Anticipatory Environmental (Hi)Stories from Antiquity to the Anthropocene.* Lanham, MD: Lexington Books, 2023.

Scott, John A. "Dogs in Homer." *Classical Weekly* 41 (1948): 226–28.

Scully, Stephen. "Reading the Shield of Achilles: Terror, Anger, Delight." *Harvard Studies in Classical Philology* 101 (2003): 29–47.

Sedia, Adam. "Shelley's 'Ozymandias' and the Immortality of Art." *The Imaginative Conservative,* November 27, 2020. https://theimaginativeconservative.org/2020/11/shelley-ozymandias-immortality-art-adam-sedia.html.

Servius. *Servii grammatici qui feruntur in Vergilii carmina commentarii,* ed. Georgius Thilo and Hermann Hagen. 3 vols. Leipzig: Teubner, 1881–1902.

Shay, Jonathan. *Achilles in Vietnam: Combat Trauma and the Undoing of Character.* New York: Atheneum, 1994.

Shelley, Percy Bysshe. *The Complete Poetical Works of Shelley: Including Materials*

Never before Printed in Any Edition of the Poems, ed. Thomas Hutchinson. Oxford: Oxford University Press, 1904.

———. *The Poetical Works of Percy Bysshe Shelley*. London: E. Moxon, 1840.

Sherratt, Andrew. "The Secondary Exploitation of Animals in the Old World." *World Archaeology* 15 (1983): 90–104.

Sherratt, E. S. "'Reading the Text': Archaeology and the Homeric Question." *Antiquity* 64 (1990): 807–24.

Shipton, Matthew. *Politics of Youth in Greek Tragedy*. London: Bloomsbury, 2018.

Sidonius Apollinaris. *Carmina*. In *Sidonius. Poems. Letters: Books 1–2*, trans. W. B. Anderson. Loeb Classical Library 296. Cambridge, MA: Harvard University Press, 1936.

Siewers, Alfred Kentigern. "The Ecopoetics of Creation." Pp. 45–78 in Alfred Kentigern Siewers, ed., *Re-Imagining Nature*. Lewisburg, PA: Bucknell University Press, 2014.

———. "Introduction—Song, Tree, and Spring: Environmental Meaning and Environmental Humanities." Pp. 1–41 in Alfred Kentigern Siewers, ed., *Re-Imagining Nature*. Lewisburg, PA: Bucknell University Press, 2014.

———, ed. *Re-Imagining Nature*. Lewisburg, PA: Bucknell University Press, 2014.

Silius Italicus. *Punica*, trans. J. D. Duff. 2 vols. Cambridge, MA: Harvard University Press, 1934.

Simmons, James C. "Bulwer and Vesuvius: The Topicality of *The Last Days of Pompeii*." *Nineteenth-Century Fiction* 24 (1969): 103–5.

Sivaramakrishnan, Murali. "Ecopoetics and the Literature of Ancient India." Pp. 65–79 in John Parham and Louise Hutchings Westling, eds., *A Global History of Literature and the Environment*. Cambridge: Cambridge University Press, 2016.

Smith, Bruce D., and Melinda Zeder. "The Onset of the Anthropocene." *Anthropocene* 4 (2013): 8–13.

Sokolon, Marlene K. "The *Iliad*: A Song of Political Protest." *New Political Science* 30 (2008): 49–66.

Solaki-Yannopoulou, Eleni. "The Symbolic Significance of the Terracottas from the Mycenaean Sanctuary at Ayios Konstantinos, Methana." Pp. 49–57 in Eva Alram-Stern, Eva, Fritz Blakolmer, Sigrid Deger-Jalkotzy, Robert Laffineur, and Jörg Weilhartner, eds., *Metaphysis: Ritual, Myth and Symbolism in the Aegean Bronze Age*. Leuven, BE: Peeters, 2016.

Solmsen, Friedrich. "Ilias Σ 535–540." *Hermes* 93 (1965): 1–2.

Solovieva, Sergey L. Olga N. Solovieva, Chan N. Go, Khen S. Kim, and Nikolay A. Shchetnikov. *Tsunamis in the Mediterranean Sea, 2000 BC to 2000 AD*. Kluwer: Dordrecht, 2000.

Sophocles, *Philoctetes*. In Sophocles, *Antigone. The Women of Trachis. Philoctetes. Oedipus at Colonus*, ed. and trans. Hugh Lloyd-Jones. Loeb Classical Library 21. Cambridge, MA: Harvard University Press, 1994.

Stallings, Alicia. "Crown Shyness." *Sewanee Review* 131 (2023).

Stanford, William Bedell. *Greek Metaphor.* Oxford: Blackwell, 1936.

Statius. *Thebaid,* ed. and trans. D. R. Shackleton Bailey. 2 vols. Cambridge, MA: Harvard University Press, 2004.

Stax. "The Stax Report: Script Review of *Troy.*" *IGN,* November 14, 2022. https://www.ign.com/articles/2002/11/14/the-stax-report-script-review-of-troy.

Stead, Henry. "The Only Tone for Terror." Pp. 205–9 in Edith Hall, ed., *New Light on Tony Harrison.* Oxford: Oxford University Press, 2019.

Stein, Charles David. "Beyond the Generation of Leaves: The Imagery of Trees and Human Life in Homer." Ph.D. diss., UCLA, 2013.

Stephens, Frederic George. "Fine Arts. Royal Academy." *The Athenaeum* 2118 (1868): 768–69.

Stiros, S. C. "Earthquakes." Pp. 469–91 in Jamie Woodward, ed., *The Physical Geography of the Mediterranean.* Oxford: Oxford University Press, 2009.

———. "Social and Historical Impacts of Earthquake-Related Sea-Level Changes on Ancient (Prehistoric to Roman) Sites." *Zeitschrift für Geomorphologie* suppl. 137 (2005): 469–91.

Stone, Christopher. *Should Trees Have Standing? Law, Morality, and the Environment.* 3rd ed. Oxford: Oxford University Press, 2010.

Storey, Ian C., trans. *Fragments of Old Comedy,* vol. 2: *Diopeithes to Pherecrates.* Loeb Classical Library 514. Cambridge, MA: Harvard University Press, 2011.

Strabo. *Geography,* trans. Horace Leonard Jones. 7 vols. Loeb Classical Library. Cambridge, MA: Harvard University Press, 1917–1932.

Subramanian, Meera. "Humans versus Earth: The Quest to Define the Anthropocene." *Nature* 572 (2019): 168–70.

Summers, Claude J. "'Or One Could Weep Because Another Wept': The Counterplot of Auden's 'The Shield of Achilles.'" *Journal of English and Germanic Philology* 83, no. 2 (1984): 214–32.

Szerszynski, Bronislaw. "The End of the End of Nature: The Anthropocene and the Fate of the Human." *Oxford Literary Review* 34 (2012): 165–84.

Taplin, Oliver. *Greek Fire.* London: Cape, 1989.

———. "The Shield of Achilles within the 'Iliad.'" *Greece & Rome* 27, no. 1 (1980): 1–21.

Tate, Andrew. *Apocalyptic Fiction.* London: Bloomsbury, 2017.

Taylour, William D., and R. Janko. *Ayios Stephanos: Excavations at a Bronze Age and Medieval Settlement in Southern Laconia,* suppl. 44. London: British School at Athens, 2008.

Thein, Karel. *Ecphrastic Shields in Graeco-Roman Literature: The World's Forge.* Abingdon, UK: Routledge, 2022.

Theophrastus. *De Causis Plantarum,* ed. and trans. Benedict Einarson and George K. K. Link. 3 vols. Loeb Classical Library. Cambridge, MA: Harvard University Press, 1976–1990.

Thirgood, J. V. *Man and the Mediterranean Forest: A History of Resource Depletion.* London: Academic Press, 1981.

Thomas, Roy, Paolo Rivera, and Miguel Sepulveda. *Marvel Illustrated: The "Iliad."* New York: Marvel Comics, 2007.

Thommen, Lukas. *An Environmental History of Ancient Greece and Rome.* Rev. ed. Cambridge: Cambridge University Press, 2012.

Thornber, Karen. "Environments of Early Chinese and Japanese Literatures." Pp. 37–51 in John Parham and Louise Hutchings Westling, eds., *A Global History of Literature and the Environment.* Cambridge: Cambridge University Press, 2016.

Thorndycraft, Varyl R., Gerado Benito, Mariano Barriendos, and M. Carmen Llasat. *Palaeofloods.* Madrid: CSIC, 2002.

Thornes, John. "Land Degradation." Pp. 563–82 in Jamie Woodward, ed., *The Physical Geography of the Mediterranean.* Oxford: Oxford University Press, 2009.

Thucydides. *The History of the Peloponnesian War.* Rev. ed. trans. Rex Warner, with introduction by M. I. Finley. Harmondsworth, UK: Penguin Classics, 1954.

Tlili, Sarra. "I Invoke God, Therefore I Am: Creation's Spirituality and Its Ecologic Impact in Islamic Texts." Pp. 107–22 in John Parham and Louise Hutchings Westling, eds., *A Global History of Literature and the Environment.* Cambridge: Cambridge University Press, 2016.

Tobin, Greg. *The Siege of Troy.* New York: St. Martin's, 2004.

Tomaselli, Ruggero. "The Degradation of the Mediterranean Maquis." *Ambio* 6, no. 6 (1977): 356–62.

Trapp, Richard L. "Ajax in the 'Iliad.'" *Classical Journal* 56 (1961): 271–75.

Trépanier, Simon. "Early Greek Theology: God as Nature and Natural God." Pp. 273–317 in Jan N. Bremmer and Andrew Erskine, eds., *The Gods of Ancient Greece.* Edinburgh: Edinburgh University Press, 2010.

Tsountas, Chrēstos, and James Irving Manatt. *The Mycenaean Age: A Study of the Monuments and Culture of Pre-Homeric Greece.* Boston: Houghton Mifflin, 1897.

Turner, Victor. *Dramas, Fields, and Metaphors: Symbolic Action in Human Society.* Ithaca, NY: Cornell University Press, 1974.

Ukrainka, Lesia. *Cassandra: A Dramatic Poem,* trans. Nina Murray. Cambridge, MA: Harvard University Press, 2025.

UNICEF. "'Barely a Drop to Drink': Children in the Gaza Strip Do Not Access 90 Per Cent of Their Normal Water Use." December 20, 2023. https://www.unicef.org/press-releases/barely-drop-drink-children-gaza-strip-do-not-access-90-cent-their-normal-water-use.

United Nations. *The State of the World's Forests: Forests, Biodiversity and People.* UN Environment Programme, May 22, 2020. https://www.unep.org/resources/state-worlds-forests-forests-biodiversity-and-people.

Usher, Mark D. *Plato's Pigs and Other Ruminations: Ancient Guides to Living with Nature.* Cambridge: Cambridge University Press, 2020.

Uysal, İsmet. "An Overview of Plant Diversity of Kazdagi (Mt. Ida) Forest National Park, Turkey." *Journal of Environmental Biology* 31 (2010): 141–47.

van der Mije, Sebastiaan R. "Bad Herbs—The Snake Simile in *Iliad 22*." *Mnemosyne* 64 (2011): 359–82.

van der Valk, Marchinus. *Commentarii ad Homeri Iliadem pertinentes ad fidem Codicis Laurentiani editi.* 5 vols. Leiden: Brill, 1971–1987.

Vandiver, Elizabeth. *Stand in the Trench, Achilles: Classical Reception in British Poetry of the Great War.* Oxford: Oxford University Press, 2010.

van Thiel, Helmut, ed. *Aristarch, Aristophanes Byzantios, Demetrios Ixion, Zenodot: Fragmente zur Ilias gesammelt, neu herausgegeben und kommentiert.* Berlin: de Gruyter, 2014.

Veldhuijzen, H. A. "Just a Few Rusty Bits: The Innovation of Iron in the Eastern Mediterranean in the 2nd and 1st Millennia BC." Pp. 237–50 in V. Kassianadou and G. Papsavvas, *Eastern Mediterranean Metallurgy in the Second Millennium BC.* Oxford: Oxbow, 2012.

Veldsman, Daniël P. "The Place of Metaphysics in the Science-Religion Debate." *HTS Teologiese Studies* 73 (2017): 1–7.

Ventris, Michael, and John Chadwick. *Documents in Mycenaean Greek.* Cambridge: Cambridge University Press, 1956.

Vetters, Melissa. "All the Same Yet Not Identical? Mycenaean Terracotta Figurines in Context." Pp. 37–48 in Eva Alram-Stern, Fritz Blakolmer, Sigrid Deger-Jalkotzy, Robert Laffineur, and Jörg Weilhartner, eds., *Metaphysis: Ritual, Myth and Symbolism in the Aegean Bronze Age.* Leuven, BE: Peeters, 2016.

Vidal-Naquet, Pierre. *The Atlantis Story: A Short History of Plato's Myth.* Exeter: University of Exeter Press, 2007.

———. "Oedipus in Vicenza and in Paris: Two Turning Points in the History of Oedipus." Pp. 361–80 in Jean-Pierre Vernant and Pierre Vidal-Naquet, *Myth and Tragedy in Ancient Greece*, trans. Janet Lloyd. New York: Zone Books, 1988.

Virgil. *Georgics.* In Virgil, *Eclogues. Georgics. Aeneid: Books 1–6*, trans. H. Rushton Fairclough, rev. G. P. Goold. Loeb Classical Library 63. Cambridge, MA: Harvard University Press, 1916.

Wahl, Jean. "Préface." Pp. 7–12 in Rachel Bespaloff, *De l'Iliade.* New York: Brentano's, 1943.

Walcott, Derek. *Epitaph for the Young: XII Cantos.* Bridgetown: Barbados Advocate Co., 1949.

———. *Omeros.* London: Faber & Faber, 1990.

Wallace, Nathaniel. "Cultural Process in the 'Iliad' 18:478–608, 19:373–80 ('Shield of Achilles') and Exodus 25:1–40:38 ('Ark of the Covenant')." *College Literature* 35 (2008): 55–74.

Wallace-Wells, David. "The Uninhabitable Earth." *New York Magazine*, July 2017. https://nymag.com/intelligencer/2017/07/climate-change-earth-too-hot-for -humans-annotated.html.

Walløe, Lars. "Was the Disruption of the Mycenaean World Caused by Repeated Epidemics of Bubonic Plague?" *Opuscula Atheniensia* 24 (1999): 121–26.

Walsh, David. "Warrior and Anti-Warrior." World Socialist Website, June 19, 2004. https://www.wsws.org/en/articles/2004/06/troy-j19.html.

Walter, Justine. "Poseidon's Wrath and the End of Helike: Notions about the Anthropogenic Character of Disasters in Antiquity." Pp. 31–43 in Christopher Schliephake, ed., *Ecocriticism, Ecology, and the Cultures of Antiquity*. Lanham, MD: Lexington Books, 2017.

Ward, William, Martha Sharp Joukowsky, and Paul Åström. *The Crisis Years: The 12th Century BC from beyond the Danube to the Tigris*. Dubuque, IA: Kendall/ Hunt, 1992.

Warner, Marina, Thomas Rayfiel, Sarah Deming, Robert Pinsky, Erik Tarloff, Anne Wagner, Arthur Lubow, and Mark Morris. "A Symposium on Crying." *The Threepenny Review* 147 (2016): 18–21.

Washizuka, Naho. "Auden as an Example: Mid-to-Late Seamus Heaney." *Journal of Irish Studies* 29 (2014): 9–18.

Weil, Simone. *Simone Weil's "The Iliad"; or, The Poem of Force: A Critical Edition*, trans. and ed. James P. Holoka. Oxford: P. Lang, 2003.

Weil, Simone, and Rachel Bespaloff. *War and the "Iliad."* Trans. Mary McCarthy. New York: New York Review Books, 2005.

Wenders, Wim. *Der Himmel über Berlin* (The heaven/sky over Berlin). Roadmovies/ Filmproduktion/Argos Films, 1987.

Wertime, Theodore A. "The Beginnings of Metallurgy: A New Look." *Science* 182, no. 4115 (1973): 875–87.

West, Martin L. *The East Face of Helicon: West Asiatic Elements in Greek Poetry and Myth*. Oxford: Clarendon, 1997.

———. *Greek Epic Fragments*. Ed. and trans. Martin L. West. Cambridge, MA: Harvard University Press, 2003.

———. *Homeric Hymns. Homeric Apocrypha. Lives of Homer.* Ed. and trans. Martin L. West. Cambridge, MA: Harvard University Press, 2003.

———. "The Homeric Question Today." *Proceedings of the American Philosophical Society* 155 (2011): 383–93.

———. *Indo-European Poetry and Myth*. Oxford: Oxford University Press, 2007.

———. "The Rise of the Greek Epic." *Journal of Hellenic Studies* 108 (1988): 151–72.

Whenzou News. "Water on Fire: Polluted Meiyu River Ignites in Whenzou China." Strange Sounds, March 15, 2014. https://strangesounds.org/2014/03/water -on-fire-polluted-meiyu-river-ignites-in-whenzou-china.html.

White, Lynn, Jr. "The Historical Roots of Our Ecological Crisis." Pp. 18–30 in Ian G. Barbour, ed., *Western Man and Environmental Ethics*. Reading, MA: Addison-Wesley, 1973.

Wilkie, Rob. "Introduction: After the Law of Value Is 'Blown Apart': Labor as Value in the Contemporary." *Minnesota Review* 87 (2016): 110–15.

Willcock, Malcom M., ed. *The "Iliad" of Homer: Books XIII-XXIV.* Basingstoke: St Martin's, 1984.

Williams, Michael. *Deforesting the Earth: From Prehistory to Global Crisis.* Chicago: Chicago University Press, 2003.

Williams, Raymond. *The Country and the City.* London: Chatto and Windus, 1973.

———. *Marxism and Literature.* Oxford: Oxford University Press, 1977.

Wilson, Emily, trans. *The "Iliad."* New: Norton, 2023.

Wilson, John R. "The Gilgamesh Epic and the *Iliad.*" *Echos du monde classique: Classical News and Views* 30, no. 1 (1986): 25–41.

Winkler, Martin, ed. *Troy: From Homer's "Iliad" to Hollywood Epic.* Oxford: Blackwell, 2007.

Woodward, Jamie, ed. *The Physical Geography of the Mediterranean.* Oxford: Oxford University Press, 2009.

Wright, Elizabeth Mary. *The Life of Joseph Wright.* London: Oxford University Press, 1932.

Wyles, Rosie. "Ménage's Learned Ladies: Anne Dacier (1647–1720) and Anna Maria van Schurman (1607–1678)." Pp. 61–77 in Rosie Wyles and Edith Hall, eds., *Women Classical Scholars: Unsealing the Fountain from the Renaissance to Jacqueline de Romilly.* Oxford: Oxford University Press, 2016.

Wyles, Rosie, and Edith Hall, eds. *Women Classical Scholars: Unsealing the Fountain from the Renaissance to Jacqueline de Romilly.* Oxford: Oxford University Press, 2016.

Index

Page numbers in italics refer to figures.